Wireless Writing in the Age of Marconi

Electronic Mediations

Katherine Hayles, Mark Poster, and Samuel Weber, Series Editors

Electronic Mediations, Volume 16

WIRELESS WRITING
IN THE AGE OF MARCONI

Timothy C. Campbell

University of Minnesota Press
Minneapolis • London

The University of Minnesota Press gratefully acknowledges the generous assistance provided for the publication of this book by the Hull Memorial Publication Fund of Cornell University.

Excerpts from "Cantos I," "Canto XVIII," and "Canto CX" by Ezra Pound are from *The Cantos of Ezra Pound*. Copyright 1934, 1937, 1940, 1948, 1956, 1959, 1962, 1963, 1966, and 1968 by Ezra Pound. Reprinted by permission of New Directions Publishing Corporation.

Sections of chapter 2 were first published as "The *Immaginazione Senza Fili* and the Noise of Modern Life: Towards a Media Ecology of Futurism," *Forum Italicum* 37, no. 2 (Fall 2003): 371–89. Reprinted by permission of Forum Italicum.

Every effort was made to obtain permission to reproduce material used in this book. If any proper acknowledgment has not been made, we encourage copyright holders to notify us.

Published by the University of Minnesota Press
111 Third Avenue South, Suite 290
Minneapolis, MN 55401-2520
http://www.upress.umn.edu

Library of Congress Cataloging-in-Publication Data

Campbell, Timothy C.
    Wireless writing in the age of Marconi / Timothy C. Campbell.
        p.   cm. — (Electronic mediations ; v. 16)
    Includes bibliographical references and index.
    ISBN 13: 978-0-8166-4441-4 (hc) — ISBN 13: 978-0-8166-4442-1 (pb)
    ISBN 10: 0-8166-4441-1 (alk. paper) — ISBN 10: 0-8166-4442-X (pbk. : alk. paper)
    1. Radio authorship. 2. Radio broadcasting—Italy—History. 3. Pound, Ezra, 1885–1972—History and criticism. 4. Marinetti, Filippo Tommaso, 1876–1944—History and criticism. 5. D'Annunzio, Gabriele, 1863–1938—History and criticism. 6. Telegraph, Wireless—History. 7. Radio—History. I. Title. II. Series.
    PN1991.7.C32   2006
    808'.066791—dc22

                                                                        2005034419

Printed in the United States of America on acid-free paper

The University of Minnesota is an equal-opportunity educator and employer.

12 11 10 09 08 07 06   10 9 8 7 6 5 4 3 2 1

Contents

Acknowledgments

OF ALL THE COLLABORATIONS, major and minor, that made this book possible, none was more vital than the one I began with Barbara Spackman in 1996. Her steadfast support of the project and her understanding of its interdisciplinary parameters (even before my own) contributed enormously to the study's cultural and literary itinerary. She has been the book's ideal reader, thus my heartfelt thanks to her. I also wish to thank Ursula Heise, whose introduction to media ecology and systems theory assisted me greatly in the later chapters on Ezra Pound. Her suggestions, especially those of a methodological nature, made me focus on the stakes involved in writing a history of the wireless. Finally I wish to express my deepest gratitude to Mitchell Greenberg for his friendship, words of encouragement, and enthusiasm for the project. That I was able to complete it is in no small measure due to his support of my own work and of Italian studies at Cornell.

Many others deserve thanks. On this side of the Atlantic, I want to thank the late Robert Dombroski, Harro Müller, Nelson Moe, and George Stade, who were involved in the very early stages of the project's elaboration. Additionally, I want to thank John Baldwin, Cal Donly, Greg Gilfeather, Ellen Philip, Daniel Rosenthal, Jim Steintrager, and Tim Ward, whose enthusiasm I was lucky to have when mine was waning. I also express appreciation to my family, who continually inspired me. At Cornell, I was fortunate to have the support of numerous colleagues; I am grateful to Maria Antonia Garces, Richard Klein, Tracy McNulty, Tim Murray, Edmundo Paz-Soldan, Michael Steinberg, Suzanne Stewart-Steinberg, and Marie-Claire Vallois for their guidance and consideration. Thank you.

On the other side of the Atlantic, I wish to thank Trevor Wright, information officer of the Marconi Corporation, and Barbara Valloti of La Fondazione Guglielmo Marconi for their assistance and generosity in locating images. Antonio Faeti and Remo Ceserani at the University of Bologna were both delightful and immeasurably helpful in navigating the notoriously byzantine Italian archive. The Istituto per le Scienze Religiose in Bologna also assisted me tremendously: its eclectic collection and always friendly staff eased the real difficulties of researching abroad.

On a more personal note, I wish to express my deepest appreciation to Bruno and Giovanna Baraldi, Kissi Cereghini, Aldo Oppizzi, and Richard, Theresa, and Nathan Rice. Their hospitality and friendship over what surely seemed like an interminably long gestation period were immensely appreciated.

I express my gratitude to the Wollemborg family for the fellowship I was awarded from Columbia University, which allowed me to travel to Bologna in 1996 to commence research on Marconi and the wireless. I also want to thank the participants of Cornell's Italian Studies Colloquium for their valuable suggestions on an early version of chapter 2. At the University of Minnesota Press, I was fortunate to work initially with Eric Lundgren; my thanks to him and the entire staff at Minnesota.

Most of all, I thank Michela Baraldi. Her loving companionship, faith in the project, and patience while I completed it *(punto e basta)* made all the difference.

WIRELESS WRITING AND THE PITFALLS OF "RADIO THEORY"

T HE READER coming to an introduction on wireless writing may be surprised by the initial distinction between the wireless and radio. Is not the term, one might ask, still synonymous in certain parts of the world with radio, to denote voice transmissions without wires? Why the need—if indeed it is a need, a question to which we will want to return—to split hairs when both terms work equally well in marking the characteristics of a well-known twentieth-century technology? A less sympathetic reader might even point to its absence in Marshall McLuhan's *Understanding Media,* recalling the "hot" entry on radio, in which he disparages the term as the communications equivalent of "the negative 'horseless-carriage' attitude toward a new form [radio]," suggesting that while everyone was busy taking the wireless for a telegraph, they could not see it "even in relation to the telephone."[1] The term apparently falls short on two counts since it fails to signal decisively features different from radio while connoting something like a premodern and negative view of technology.

Yet the development of wireless technology in the past decade, even during the past five years, belies McLuhan's judgment. The growth of wireless telephony and the heightened connectivity of wireless networks are dramatically altering the role assigned to wireless technology in the present media ecology. No longer a quaint English term for radio, the wireless is now is a mode of communication involving the Internet, e-mail, telephony, computer hardware, and, of course, radio. The "wireless" has seemingly lost its negative connotations, as it highlights qualities specific to the technology, namely, its capacity for transmitting between points without the aid of wires. Oddly though, greater awareness of wireless communication among the general public as a means of communication has not quite led to anything like a reappraisal of wireless technology among historians and cultural critics. In most accounts, it continues to play the well-respected if dowdy grandfather to radio's well-traveled young man. Although I do not have anything like the exact figures at hand, studies of radio in all its various contexts (cultural, social, political, psychological) continue apace while those examining the wireless as a distinct entry in a media lexicon are few indeed. This is not to say that Marconi and wireless invention do not make an appearance in the abundant historiography of radio, but more often than not the wireless inhabits the gray zone between telegraphy and humble genealogies of early radio; its true significance is linked either to telegraphy as its culminating moment or radio as its privileged (and forgotten) origin.

One finds much the same marginalization of the wireless in contemporary cultural and media studies, especially among those interested in thinking the relation between modern cultural and literary production and its medialogical condition. The reason perhaps lies with McLuhan himself, judging from the kinds of analyses offered over the past three decades. His dismissal of the wireless continues to haunt recent approaches to radio culture, especially when the investigation turns to some of the principal figures of heroic modernism across Europe: Pound definitely had his "devil-box" and wrote opera for the radio; Benjamin and Orwell both wrote for radio and Gadda and Marinetti spoke over it in Italy, despite the latter's transmission failures in the 1930s.[2] But in at least two of these cases (Pound and Marinetti), their encounters with the wireless began twenty years before radio, a fact that demands our attention. Indeed, I know of no study that posits the wireless as a distinct communications medium in the first quarter of the past century or that examines its autonomy vis-à-vis telegraphy, radio, and telephony or its impact on modern literature. The results have been unfortunate: differences between the wireless and radio are elided, or, more problematically, characteristics more properly belonging to the radio are evoked to mark early modern textual forms that owe if not their genesis then a certain inflection to wireless invention.

The following study is an attempt to think the relation between changes in communication media brought on by the wireless with certain forms of modern literature. Its historical frame is slightly different from studies of radio in that I begin ten years earlier in 1880s Italy with a series of experiments that measured acoustic spacing and end with the coupling of wireless to gramophone and sound storage that emerged in the early 1930s. The emphasis on spacing, sound, and systems of media does not occur by chance since these categories figure prominently in the distinction I wish to draw between radio and wireless or, more accurately, between an interpretive framework based on voice transmission and one based on the inscription of sounds. In the sections that follow, I want to spell out the principal components of what I call radio theory and then detail the chief features of wireless writing. For the former, this includes an overemphasis on the spoken qualities of transmission and a blindness to the role storage devices play in capturing voices and sounds for later transmission. For the wireless, this consists in a staggering capacity for coupling with other media, including those of the body, and in the information processing of sounds that it gives rise to. Taking up the optic of wireless writing allows us, I argue, to draw more sustained and productive relations between literature and communication media of the period while offering the possibility of comprehending and narrating important shifts in modern textual forms and their concretization through wireless media.

Radio Histories

Since some readers may be unfamiliar with the details of early wireless invention, I want to begin by elaborating the technical distinction between wireless and radio. Wireless technology refers to the electrical signaling of messages without the aid of cables. The combination of electricity and the lack of wires are key as they distinguish the wireless of the last century from its nineteenth-century predecessors: the optical telegraphy of the Chappe system in operation in France between 1794 and 1855 and the Morse telegraph of midcentury America. When Marconi first introduced his wireless in 1895, it was essentially a variation of the telegraph without the wires (captured in the Italian *telegrafia senza fili* and the German *die drahtlose Telegraphie*), which relayed Morse code signals tapped at a transmitter to a receiver. Marconi was aided crucially by James Clerk Maxwell's theory of electromagnetic radiation published in 1865 and by Heinrich Hertz's experimental proof of the existence of radio waves in 1888, better known as Hertzian waves. Marconi's achievement was to see in these early radio waves the possibility for sending messages (marconigrams) using Morse code.[3] Improving on the work of others (when not appropriating them outright), especially from Oliver Lodge, Edouard Branly, and Nikola Tesla, Marconi utilized primitive circuits and galvanic batteries to set up "spark-gap transmitters" that sent very weak electromagnetic waves to receivers that consisted of iron filings.[4] These iron filings "cohered" or "discohered" depending on the reception of the Hertzian wave, thus marking the dots and dashes of Morse code. Marconi was equally noted for his antenna designs, his formulations for arranging circuits, and the introduction of the ground-plane, or grounded vertical antenna, for wireless transmission and reception. Of singular importance was Marconi's discovery that long waves followed the curvature of the earth; in practice this meant that the Marconi receiving unit was not required to be in the transmitter's line of sight. Thus Marconi was the first to transmit wirelessly across, over, and around all obstacles simply by using the earth.[5] Despite these advances, the device was limited by the weakness of its signals until a method could be found to amplify them at the receiving end or to increase the power of the signal at the site of transmission. Transmitting Morse code was one thing; regular voice transmission was another given the greater power needs required.

The successful resolution of these problems took another twenty years and a world war to overcome. Although I describe this history in some detail in chapter 4, it may be summarized here. Beginning with the construction of high-frequency alternators before and during the war (the Alexanderson and Goldschmidt generators in the United States and Germany, respectively), the power of the wireless signal was increased, allowing for wireless voice transmissions to occur if not routinely then more frequently. More significant was Edwin Armstrong's development of the superheterodyne, a device originally intended to pick up the high-frequency transmissions the Allies feared the Central Powers were using to communicate with their biplanes. That they were not did not alter the magnitude of Armstrong's invention. Essentially, the superheterodyne allowed listeners to fine-tune frequencies by heterodyning (from the Greek *hetero*, "other," and *dyne*, "to force together") particularly weak signals or inaudible frequencies, resulting in its audibility in the sound of a third frequency. To produce the third signal, the wireless came equipped by 1925 with a wave-producing circuit that could generate a second frequency to be heterodyned with the first. What was unheard or inaudible was now only a superheterodyne or turn of the dial away.

The final wireless advance in the period being discussed occurred in the late 1920s and early 1930s and concerned the heightened interconnectivity of wireless transmitters with the storage media of the gramophone and sound strip. Although the transmission of gramophonic records via wireless had begun in a limited fashion some years earlier, it was beset by discrepancies between the range of frequencies a wireless utilized and those that could be inscribed on a record.[6] Once the frequencies in voice transmissions and technological storage converged, sounds could be cut and mixed in montage, resulting in important temporal effects, especially in the field of time manipulation. According to an early student of the wireless, sound montage made the listener "independent of the time and the place of production," because "original records of real sounds as well as of historical events can be mingled."[7] The transmission of recorded sounds generates an organic media network of equal sound values that combines the acoustic data of stored sounds with voices transmitted at the wireless. Accordingly, a processing of sounds was born that depended on hearing them together and against other contiguous sounds.

The Logosphere of Radio

This admittedly incomplete sketch of wireless development from its inception to its greater coupling with storage media helps identify the historical moment that radio proponents continue to privilege as the instantiation of institutional broadcasting of voices to large numbers of listeners that occurred in the late 1920s and 1930s in Europe and North America. It is radio's singular ability to create and nurture a collective audience that has contributed enormously to a modern radio imaginary and semantics. Hadley Cantril's influential study from 1935, *The Psychology of Radio,* is typical of the symbolic effects associated with the beginnings of large-scale voice transmission. Radio "fills us with a 'consciousness of kind' which at times grows into an impression of vast social unity. It is for this reason that radio is potentially more effective than print in bringing about concerted opinion and action."[8] In Cantril's estimation, listeners are souls that commune with each other across nations, times, and space, brought together by an overarching voice that forges links between them. Writing forty years later, Gaston Bachelard echoes Cantril's communing souls to construct an ethics and aesthetics of the logosphere. For Bachelard, radio founds a universal world in which "everyone can hear everyone else and we can all listen in peace." What term, he asks, is better suited to "this domain of world speech than 'logosphere,'" since we all speak from within it. We are indeed "citizens of the logosphere."[9] Much the same vision informs R. Murray Schafer's notion of an acoustic community, with vital information reaching individual subjects in the intimacy of their homes in order to create an ideal group of listeners.[10] That a radio imaginary is still in operation today may be judged by the market niche occupied by cultural studies of radio and their attention to the ways radio shaped imagined communities as well as its relationship to the construction of a private and collective sphere of individual listeners.[11]

Foregrounding voice transmissions in radio theory is problematic, however. First, it assumes something like a logocentric notion of speaker and listener, which Bachelard's logosphere elegantly captures: we suppose that the effects of text are reducible to the meaning intended by the speaker (and the voice), and to "even its supposedly unique and identifiable signatory," the listener.[12] The ease with which many histories associate radio with intimacy and the private world of listening is largely symptomatic of such a bias; radio theory posits a relation between the creation of interiority and the obligatory reception of the transmission.[13] The result is a notion of subjectivity that shares many features with print culture's depth model of subjectivity.[14] In radio, the acoustic world of the lone voice reinforces the interiority and individuality of the speaking subject so that one fails to hear the voice as anything other than an immaterial verbal construction in which space collapses "to an ideal of instantaneous transmission and reception, a communication without mediation."[15] The speaker is "present" to his listeners to the degree the transmission of voice creates the illusion of resemblance.

Second, by overlooking the series of couplings with storage media that made possible the retransmission of voice, radio proponents fail to observe what John Johnston, following German media theorist Friedrich Kittler, terms "partially connected media systems," or a communications assemblage in which the "diversity of information and effects" produced by a period's technologies become the conditions underlying a literary text's genesis.[16] Thus radio's mediality would not include simply the microphone and its amplification of voice but also the possibilities for the inscription of voice onto disk or in (type)writing, as well as its later retrieval and transmission. In other words, radio as a theoretical framework does not adequately mark the connections running to other media, including bodily media, in what we today call an interface.[17] The blindness of radio proponents to partially connected media systems might be dismissed if our attention were to fall merely on voice amplification and the soundscape of radio worlds. Yet, when our uppermost concern is narrating the shifts in communication media that modern literature registers, we require an optic that points us in the direction of readings that can narrate the varying ways media and literature collaborate. Readings in the radio key are today of limited help in theorizing the relation between the two, which, I argue, is especially the case with F. T. Marinetti's Futurist manifestos and Ezra Pound's *Cantos*.

Wireless Inscription

Thus for the media theorist interested in systemic relations among media across an extended historical period, the term radio lacks specificity. The wireless, as I deploy the term here, is more precise to the degree it names not only the telegraph Marconi invented in 1895 but also the series of couplings and interfaces over a thirty-year arc. Before its partial connection with voice storage on discs, the wireless collaborated with other media: in its earliest format, the wireless operator, or *marconista*, listened to signals that emerged out of the static of his headset and then wrote them down. The writing down of sounds was possible thanks to a relay running from the operator's ear to his or her hand, which utilized the storage possibilities of the alphabet itself. With the increased use of the typewriter that accompanied communication practices of World War I, a second moment of wireless connectivity emerges: now the typewriter partially mechanizes the function of the human interface on the wireless, spatializing a general stream of data in order to prepare it for its transmission into a binary of dots and dashes. Following Kittler's trajectory of media connectivity in modernity, the final moment of media collaboration occurs more or less in the early 1930s when it becomes possible to store, retrieve, and transmit previously archived voices and sounds over the wireless. Although the connections in the media system undergo significant changes over the course of thirty years, what remains stable is wireless connectivity and its capacity for interaction with other media, bodily and other. Given its storied faculty for connecting with other media, wireless growth in the present media ecology will perhaps seem less surprising.

Yet the wireless is not simply a yardstick for measuring technological proclivities for heightened connectivity with other media (which the term "wireless" brilliantly captures). As I suggested above, a second feature concerns the fine-tuning of signals across the broadcast band, which the introduction of the superheterodyne made possible. Whereas early wireless transmissions registered in writing had difficulty switching frequencies and hence capturing voice, the introduction of the superheterodyne meant that a wireless operator could fine-tune numerous stations, even those in close frequency. Where the ear could not distinguish between near transmissions in the previous format, the superheterodyne could do so in the new format. The importance of the invention may be measured by the enormous effects it had on the European collective mentality: the ability to pick up the formerly unheard mesmerized many at the time, including one of this study's chief protagonists, Ezra Pound, and a whole cadre of fascists across Europe.

Considering this itinerary of wireless development, where does the notion of wireless writing part ways with radio theory when the object of study is the literary text? First, by outlining the interface among hands, voice, and early wireless inscription technologies, our attention will shift to texts that register data flows in a sort of wireless dictation and to those forms that mimic wireless encrypting techniques. No work more deftly embodies wireless registration than Filippo Tommaso Marinetti's, whose *parole in libertà* (words-in-freedom) are designed to capture acoustic, visual, and tactile sensations reaching bodily media. Second, when we scrutinize the linkup between the wireless and storage media, a series of recursive complexities (between speaker and listener and speaker and the self-presence of voice) emerges with significant consequences for how we will want to understand radio broadcasting. In a suggestive reading of Rainer Maria Rilke's "Primal Sound," Kittler argues that in a founding age of media, it becomes possible for students to hear their own voices, "not their words and answers as programmed feedback by the education system, but the real voice against a backdrop of pure silence or attention."[18] Kittler's emphasis on attention as the frame in which listening takes place recalls Ezra Pound's own theories on wirelessly generated correlates of attention in *Guide to Kulchur,* and in particular the opportunities offered by the transmission of his own voice in tracing grooves in listeners' minds by means of sequential manipulation. Pound's broadcasts over Radio Rome, traditionally viewed under a radio optic, may be productively read as an exercise in time manipulation when our focus turns to wireless connectivity with disk storage.

Finally, by recognizing the wireless capacity to hear the unheard thanks to the superheterodyne, we will want to be cognizant of those instances in which writing attempts to register voices, especially its attempts to modulate frequencies through typography, onomatopoeia, punctuation, and, most important, quotation marks. Here Pound's *Cantos* are central to the degree they register and switch among voices a full decade before the emergence of radio broadcasting in Europe. The wireless and not radio accelerated and condensed Pound's poetry, forcing it to develop a capacity for registering transmission on more than one frequency. Thus the term "wireless writing" designates the frequency modulation that certain literary forms, but most dramatically Pound's poetry, enact in the period immediately following the First World War.

Revealing Fascism

Beyond the heuristic advantages noted above, the term "wireless writing" is helpful in relating fascism in Europe to technology in the interwar period. Aside from the discussion of Marconi and emergent wireless practices in chapter 1, each subsequent chapter contains an extended reflection on how certain features of wireless writing become linked to fascist symbolic, somatic, and political values. I am particularly interested in the cultural matrix associated with the wireless in the period immediately preceding the fascist grab for power in 1922 and equally how the wireless became identified with an apocalyptic tone and rhetoric meant to generate alarm in its subjects. By examining the speeches Gabriele D'Annunzio gave over the eighteen-month occupation of Fiume, my analysis is directed toward linking the *marconista*'s deterritorialized perceptions and wireless protocols of writing with a number of fascist practices. The principal affinity resides in their similarly generated apocalyptic tones and the sanction the wireless provides for the commandant's and Il Duce's status as modern prophets.[19] For Marinetti, I point out how the Futurist manifestos detail procedures for recruiting a public modeled on wireless practices and acoustic spacing; in its readjustment of the definitions of operator, transmission, and writer, a new set of ground rules for wireless writing appears with significant effects for interwar political syntax.

The most sustained consideration of wireless writing and fascist theoretical conventions occurs in my discussion of Ezra Pound's notion of *kulchur* in chapter 5. By unearthing the term's debt to gramophonic inscription, I argue that both *kulchur* and fascism exploit wireless acoustic spacing to inscribe commands in a flash; with the evolution of a wireless media ecology, frequency modulation and the altered time axis of acoustic events become two of the principal media conditions of fascism. Put differently, I view both *kulchur* and fascism as wireless systems that exploit spacing to inscribe commands, isolating the route to the ear in oracles and continually deferred revelations. The essential idea is that fascism employs the operational mode of the wireless system, that is, it appropriates a sort of information processing that moves through blindness and the contiguity of superimposed sound values of the wireless-gramophone hookup. Contrary to current perspectives on fascism, the fascist in my view is not simply the fascinated user of technology, nor a god prosthetically extended by technology, but rather a body that has reached a certain threshold of speed by way of the spacing of vibrations and gramophonic inscription of those parts of the brain that can register. A robust examination of the materiality of wireless writing can help account for how certain cultural forms change under the impact of wireless technology.

Open Questions

There is a final caveat on the theoretical framework of wireless writing. I do not wish to suggest that the term as I unabashedly deploy it here accounts completely for the genesis of literary and poetic forms in the first part of the large century or more locally in the cases of D'Annunzio, Marinetti, and Ezra Pound. As N. Katherine Hayles wrote: "[L]ike scientific experiments, texts may rarely or never be completely explained by a given theory; there will always be elements that resist incorporation into a theoretical matrix."[20] Thus I have attempted throughout the study to mark those moments of resistance to incorporation, especially in the complicated nexus of relationships among fascism, wireless technology, and cultural production. Moreover, in bringing these objects of study together, I have obviously excluded others, typically art, and not only from the period in question; I fail to discuss important contemporary cultural engagements with the technology that might provide windows on wireless functionality. One wonders how, for instance, recent textual forms, be they the traditional print-based or digital, account for, reproduce, and narrate the increasing role of wireless technology for the present media ecology. How is the heightened connectivity of wireless media made contingent on culture and the symbolic order that precedes it? A number of contemporary artists are exploring these questions in their works and providing the coordinates for a future mapping of wireless media.[21] When read in conjunction with their imagined responses to the present media ecology, this study may contribute to a broad-ranging discussion on the wireless and its cultural reception in the age(s) of Marconi.

1

MARCONI, *MARCONISTA*: THE BIRTH OF RADIOTELEGRAPHY

FOR THOSE FAMILIAR WITH WIRELESS VISUALS from the past century, the cut wire is perhaps the most important distinction by which the wireless, or more precisely the "telegraph without wires," is distinguished from wired communications of the period (the cabled telegraph and telephone). Popular accounts of Guglielmo Marconi's experiments in the years 1895 to 1905 focus obsessively on the absence of wires and the effects these missing wires might have on new forms of communication. The wireless quickly becomes enmeshed in an unusual assortment of technologically inflected discourse. Mental telepathy, spiritual encounters with the dead, and the recuperation of past histories are all associated with a new communication device that seemingly could forego the preceding materialities of wired transmission and reception. Interest in the wireless was not limited, however, to readers of the *New York Times* or *Wireless World* or any of the vast assortment of journals dedicated to early radio but extended to some of the most important figures of heroic modernism across three continents: F. T. Marinetti praised his simulated wireless imagination in the Futurist manifestos; Vicente Huidobro evoked the wireless in "Tour Eiffel"; Gabriele D'Annunzio programmed his legionaries at Fiume to hear the signal that escaped the space-time constraints of the previous century; and Ezra Pound promulgated governments "by handwriting, by newspaper, by radio."[1] Each presuppose and propose new modes of communication made possible by the wireless while also registering a series of anxieties about the nature of cutting and transmissions that no longer depended on wires.

As significant as these contributions are for delineating the seemingly paradoxical materialities of wireless communication in modern literature (an interdiscursive network of practices linking the production and consumption of texts), I want to begin my examination elsewhere, in this instance with the first experiments in wireless transmission. My dramatis personae are Guglielmo Marconi, inventor and heir to the Jameson whiskey distilling fortune, the equipment that doubles him, *la telegrafia senza fili,* or wireless, and the experiments that led to the invention's emergence in the early years of the twentieth century. I have done so chiefly because these experiments eloquently distinguish the forms the wireless will take over the course of its development. With recourse to a number of classic Marconi biographies, I will be reading the wireless as a kind of information processing in the period's media ecology. In utilizing this concept, I am of course drawing on Kittler's theorization of Romanticism and Modernism in terms of their media technologies.[2] I have preferred his model for two reasons: (1) in its distinctions between storage and transmission technologies, the relation among body, media and subjectivity is explicated in ways that seem particularly relevant for the wireless; and (2) his notion that "communications technologies can no longer be related back to humans," precisely because they inscribe humans, parallels wireless practices and their effects on operators.[3] These will become clearer as I proceed, especially when I turn to the coupling of the writing hand with the listening ear and the nonhuman dictation that results.

I should state immediately that implicit in my reading of wireless effects will be the greater degree of coupling in the wireless operator when compared with his or her telegraphic predecessor. If "the genius of the new communications device [the telegraph] was that it incorporated the hands, ears, nerves, and brain of its operator, who merged with it to form a doubly empowered device for transmitting and receiving information," then the wireless will further extend technological incorporation given its ear/hand coupling.[4] Similarly, I shall show that the emerging position of the wireless in twentieth-century communicative exchanges altered the relation between the technological subject, the *marconista,* and his writing tools through the creation of more powerful inscription techniques. Instead of interpreting the string of dots and dashes spit out by the telegraph's Morse machine, the wireless operator hears a string of weak signals over a headset and determines which are meaningful. These he copies down, surrendering himself to what has already been selected, combined, and transmitted in a technologically inflected economy of signs; as I understand it, he is capable of hearing without copying and copying without understanding. A kind of programming ensues that moves through the interface of wireless operator and machine and the separation of acoustic and visual data flows that I term, following German media theorist Kittler and John Johnston's lead, "a partially connected media system."[5] By ascertaining the degree to which equipmental couplings result in technological subjects, "the intimate *coupling* of bodies and machines," I am explicitly suggesting that a mode of wireless writing develops that may be useful in marking certain forms of modern cultural production.[6]

Yet the chapter's itinerary moves beyond the wireless media ecology for I also discuss "Marconigrafia," the body of myths and lore that grew up around the boy inventor and his invention. Many of these sources—Marconi biographies, eyewitness and almost eyewitness accounts of the inventor by assorted friends, personal assistants, wives, daughters, and hangers-on—provide a wealth of detail on the origins of early radiotelegraphy, its cultural reception, and the uncommon anxiety the device provoked among many in the first half of the twentieth century.[7] Indeed much of the ensuing discussion might be described both as an abridged tour of Marconigrafia and an examination of the conflicted cultural response to the wireless in Italy, the United States, and Great Britain. Marking their own anxiety toward the wireless and the forms of modernity with which it is associated, these accounts provide an imaginary mapping of the effects of wireless communication and so deserve close scrutiny.

The chapter is divided accordingly into two sections. In the first I examine the classic experiments that spell the birth of the Marconi wireless and elaborate their chief similarities in order to designate emerging practices associated with a wireless media ecology. In the second, I turn more formally to Marconigrafia, to registration of anxiety unleashed by the device, and to the rhetoric deployed to counteract it. Put otherwise, I ask how the Marconi biographer writes about the wireless without deliriously compounding the biography's destination. My assumption throughout the chapter is that the Marconigrafia collected here reproduce the uncanny nature of the wireless in narratives of anxiety and desire as well as offer experiential accounts of transmission, reception, and inscription over the wireless. While some may argue that such a method risks blurring the lines between imaginary responses and the "real" operations of the wireless/ human interface, I believe a contrapuntal reading is one of the few avenues open to an observer of early radiotelegraphy wishing to pinpoint the changes the wireless enacts on perception and corporality. As the reader will soon discover, the two are intertwined to great effect in the Marconigrafia.

Preliminary Tests

Gunshots and the Postal Principle

Before examining the series of classic experiments that spell the birth of early radiotelegraphy, a preliminary word on Marconi's equipment and its operation is in order. The initial Marconi wireless, as it was called, resembled nothing more than a small and fragile glass tube about the thickness of a thermometer and two inches long. Marconi had incorporated a number of different electrical pieces in the initial design: an induction coil as a Hertzian wave–emitter and four brass balls immersed in Vaseline oil, separated by small gaps. A telegraph key was "wired" into the primary circuit of the induction coil, enabling Marconi "to cause sparks to leap the gap in dot-dash form, corresponding, of course, to the length of time the key was held down."[8] When Marconi connected one terminal of the spark discharger to a metal cylinder on top of a pole and the other to the ground, the wireless was ready to connect his transmitting station to its receiver. The elevated aerial represented an important advance as it used distance to increase signal strength. Inside the glass tube were two silver plugs "so close together that a knife blade could scarcely pass between them," with fine nickel dust in the slit. When a current passed through the slit, the particles formed a continuous metal path that fell apart under the blow of a tapper. Writing in 1937, Orrin Dunlap, Marconi's then leading American biographer, lays out the practical operation in loving detail:

> When the signals came down the antenna wire and struck the coherer, the dust particles of metal cohered (hence the name); the tapper's tiny hammer hit against the glass tube. That blow decohered the metal particles, stopping the current flow from a local battery. Each successive impulse reaching the antenna produced the same phenomenon of coherence and decoherence, hence the recording of dots and dashes on a Morse receiving instrument.[9]

Transmitter and (dis)coherer, antenna, and Morse-receiving instrument form the basis for what will be known somewhat unimaginatively as the telegraph without wires.

In Marconi lore, three experiments contribute principally to the birth of the wireless. Dunlap narrates the first experiment from 1895:

> The receiving station was carried out to a hill 1,700 meters from the window of the wireless room of Villa Grifone, so the inventor might keep an eye on the entire "expanse" he hoped to cover. At the receiver he stationed Alfonso and told him to wave a flag should he see the coherer's hammer tap three dots, the Morse letter "S." Marconi touched the telegraph key and immediately his brother's flag waved.[10]

The Marconi transmitter circa 1895. Courtesy of the Archives of the Fondazione Guglielmo Marconi, Italy.

All the components necessary for rudimentary information processing are present in the experiment: the alphabet provides a data source for the transmitted "S"; a coupling between Marconi's hand and the telegraph keys the transmission; and a coherer linked to Alfonso receives the data. By using Alfonso, Marconi was departing from a classic telegraphic model, for lacking a resistant coupling between the coherer and a Morse-receiving instrument (the Morse machine as it was known), Marconi instead utilized Alfonso as the system's pattern-recognizing component. Alfonso watches to see and hear if the coherer's hammer "taps" three times, at which point he signals Guglielmo by means of the visual medium of the flag. This convenient and ancient mode of signaling traces a line of sight from Marconi to Alfonso, who waits and watches for three dots. That Alfonso sees what Marconi has sent limits the specificity of the experiment considerably: for a device the future of which is so closely linked to the acoustic reception of weak signals, the first test is surprising for the weight afforded visual recognition. Future tests will elide the Morse machine and coherer and instead employ the human ear for its sound-registering potential.

We can also localize a further attribute of the wireless, one true of media generally. The immediate difficulty of locating the exact destination for Marconi's "S"—is it the coherer some distance away, Alfonso stationed nearby who signals his brother, or Marconi who sends himself the "S"—suggests that the wireless imposes a recognition of the other's precedence in its transmission, since no "S" could have been sent and received had Marconi not waited for the signal to return to him. While I do not wish to belabor this point, one made by Jacques Derrida concerning the post generally and Jacques Lacan in his study of "The Purloined Letter" in particular, the wireless in its early manifestations and especially in the biographies collected here is preceded by Marconi as other.[11] Briankle Chang has coined the helpful term "postal principle" to identify the rule of delivery that governs all mediations as communicative exchanges. Such a rule is clearly operating in the first wireless exchanges between Alfonso and Guglielmo. Chang writes:

> The postal principle, qua universal medium, achieves this governance by
> establishing a system of equivalences, a structure of substitutability, within
> a given network of postal relays. By providing a universal measure of system-
> wide substitution, a standard for intrasystemic translations, the postal prin-
> ciple effects the reduction of the different into the same, the domestication of
> the alien into the customary, and, as urban planners say, the gentrification of
> the unfit into the acceptable.[12]

For the reduction of difference to occur, the identity of the addresser and addressee as well as the proper functioning of the delivery mechanism must be in place. When they are, "the message (that is, that which represents the subject) in fact travels inside a closed circuit, inside a homogenous space; like a letter, it moves from one address to another without ever leaving a premapped territory."[13] Mediation in wireless communication presumes the difference between Marconi and Alfonso, which in turn presupposes their placement as identifiable points of contact in a closed circuit. Chang speaks of universal substitutability across all postal relays so that the homogeneous space of a premapped territory would be equivalent regardless of the medium in question.[14] In the space produced by the postal wireless principle, the name of Marconi is repeatedly substituted for the message and the difference between messages he delivers to himself. This is reflected in the fashion of many in Great Britain in the early part of the last century who "when there was an occasion to send a message via the telegraph without wires, simply said: 'I sent a Marconi.'"[15] Doubling of course goes hand in hand with the postal principle, and here it is Marconi doubled both for the delivery and the message received (known soon after as the marconigram). As those familiar with D'Annunzio's Fiume orations know, the doubling of Marconi for wireless transmission and reception is hardly innocent since the delivery of marconigrams is a singular component in the discursive connection set up between D'Annunzio and the legionnaires in the occupied city.

Irresistible Impulses and the Supreme Author

The second test foregrounds another property of the wireless: the signaling through, across, and around all obstacles. Conducting the test out of eyeshot, so that there be no mistake in having heard the "S" a second time, Marconi instructs Alfonso to fire a rifle when he hears the coherer tap out "a cricket-like sound." Marconi biographer Adelmo Landini describes the sound event:

> And at that moment a rifle shot reverberates beyond the celestial hills, extending to the Reno Valley. The young hero seemed to say, "Away with wires, away with heavy matter; all I need is the electric pulse for transmitting what is in the head and heart." Marconi cries out: "The wave went through, it went through, Mother, Father!" And he throws himself to his knees in an irresistible impulse of gratitude to the Supreme Author of all things. . . . And while mother and son and father embrace . . . out comes Fido like an arrow, who jumps on Guglielmo, whimpering with joy.[16]

The experiment successfully demonstrates that the wireless does not require a line of sight in order to signal: the long wave, or *fascio* as Marconi calls it, could travel over a hill and across a valley, which explains Marconi's cry, "the wave went through." Yet a less officious description would surely draw attention to the passage's odder elements. First, a gunshot is taken as a synecdoche for the larger wireless project of calming the epistemological anxiety associated with the wired materiality of the telegraph and telephone. It is not simply the passage of a signal that is celebrated but instead the end of wires and heavy matter. A new medium is born, Landini suggests, which operates without meeting the requirements of wires, foremost among them embodied transmission and reception sites, witnessed of course in Marconi's unspoken commentary that electrical impulses are enough for hearts and minds to communicate. Indeed, what is important to note here is that the signal presumably circumvents the body in transmissions, disembodying Alfonso and Marconi along the way. The disembodied *marconista*, which the Marconi narratives posit as one of the principal effects of wireless media, is soon confirmed in the irresistible impulse that overcomes Marconi, allowing him to be reached by the supreme author of all things (as well as Fido). In wireless contact, distilled spirits and not bodies communicate.[17]

The Transatlantic Paper Trail

The third test, on December 14, 1901, was designed to send a wirelessed "S" between North America and England. Accounts of the experiment foreground its similarities with previous transmissions, the clicks and eventful silence that follow the passage of a tap, while helping to determine the effects of uncoupling speech from writing in the acoustically sophisticated wireless operator who transcribes but does not speak. Marconi recounted the event to the BBC thirty years later.

> Suddenly, about half past twelve there sounded the sharp click of the "tapper" as it struck the coherer, showing me that something was coming and I listened intently. Unmistakingly, the three sharp clicks corresponding to three dots sounded in my ear; but I would not be satisfied without corroboration. "Can you hear anything, Mr. Kemp?" I said, handing the telephone to my assistant. Kemp heard the same thing as I. P. W. Paget, a little deaf, was unable to hear it, and I knew then that I had been absolutely right in my calculations. The electric waves which were being sent from Poldhu had traveled the Atlantic, serenely ignoring the curvature of the earth, which so many doubters considered would be a fatal obstacle, and they were now affecting my receiver in Newfoundland. I knew that the day on which I should be able to send full messages without wires or cables across the Atlantic was not far-distant.[18]

The first transatlantic transmission faced what would be a continuing problem, that distinguishing between code tapped by Marconi's associates at Poldhu and the noise created by atmospheric disturbances was in the best of circumstances difficult (without dwelling on the presence of the hard-of-hearing Paget). The fear was that Marconi and his assistants had quite possibly misheard the transatlantic "S."[19]

That the question was now one of mishearing at the site of reception points to a number of changes Marconi had undertaken in the wireless makeup in the intervening years. The most significant alteration was the addition of a tele-

Marconi awaiting the transatlantic signal. Reprinted by permission.

phone in order to pick out the faint "S" from the static. Landini quotes Marconi again on the reconfigurations made:

> I discovered something very simple: connecting the receiver's coherer with a simple telephone instead of a Morse machine, I can hear the sound wave even if it is very weak, while this would not have been possible with a Morse machine since it requires a rather intense sound wave in order to function.[20]

Whereas the rudimentary wireless had depended for its reception on the coupling of the coherer with the visual recognition capacity of Alfonso, Marconi now availed himself of a telephonic headset, which, in its coupling with the human ear, offered a far more sensitive receiver than the Morse machine (which required a strong signal in order for the signals to be marked on a tape). Operating the wireless now depended on the acoustic properties of the ear, the most sensitive receiver available to Marconi. I will have more to say about hearing over the wireless, but at this point it is enough to note what engineering changes resulted. On the transmission side the power requirements of the wireless decreased since earlier inscriptions via a Morse machine had required ever larger machines capable of producing sparks. Instead these generators could be used to increase the distance the signal traveled and not to meet the demands of a Morse inscription.

This is not to say Marconi completely shunted the wireless coupling with the telegraph. Its tactile and visual features remained vitally important so as to confirm that a wireless transmission had been made.

> The telegrams received were signed on every occasion by Captain Mills and the first officer, Captain Marsden or by other officials aboard the steamer. It was no longer the case of a weak sound on the telephone, but instead of points and lines registered over the telegraph, which everyone could feel, touch and check.[21]

A telegraph signs the first wireless transmissions by telephone so everyone could see, touch, and check the message. It provided, in short, a paper trail that could be used to dispel any creeping anxiety as to whether Marconi had indeed heard the "S." This represents a modification not of the identity of the ear but rather of the autonomy of acoustic data flows that reach the ear and the assuredness with which one judges how well the ear of a wireless operator can reproduce copy. Indeed, it seems likely that the authority with which the ear is invested in radiotelegraphy may owe its weight to this interaction between headset and ear. Considering that the wireless was the lone medium available for transmissions from ships to shore, the question of whether the *marconista* had heard an "S" correctly could be of grave importance; the salvage of the *Titanic* being the most spectacular example. As for the process whereby the wireless separates visual from acoustic data flows, this appears to be one with a history of autonomous data flows enacted in the partially connected media system of the gramophone, film, and typewriter described by Kittler. The transatlantic wireless separates acoustic data from the visual and tactile, furthering the autonomy of data flows outside of the previous domain of writing. The wireless, to paraphrase Kittler, arrests the acoustic data flow in signs, which will ultimately depend on what a culture recognizes as data.[22] The term wireless operator or *marconista* will name the modern functionary who aurally distinguishes the static emanating from his headset from the meaningful signs of the transmission, which he copies down.

The Birth of the *Marconista*

I want to suspend for the time being the description of the culture that grounds the recognition of data in a wireless communication system and instead focus my attention on the machinic-human interface and technological standards of the *marconista*. According to Landini, the *marconista* is to be juxtaposed not simply to the previous century's telegraph operator but also to the Morse machine itself, for the wireless operator is a hybrid of both.

Instead of interpreting the dots and dashes of the telegraphic alphabet from the Morse machine's strip of paper, now the so-called dots and dashes were listened to with a telephonic headset. After some practice, one could transcribe the meaning instantaneously on a piece of paper. At that moment the real and true wireless operator was born, the one we see today with his headset, ready to spring to another's aid, one with his apparatus, having the possibility of concentrating his hearing on a given signal and tonality and to distinguish it rationally from the atmospheric discharges and other transmissions; to perceive very weak waves augmenting consequently the distances of communication. A Morse machine could never do all that. It required a strong signal without interference and discharges of any sort in order to work. Otherwise it did not work or it would spit out a strip muddled with indecipherable dots and dashes.[23]

In compelling fashion, Landini identifies the most significant features of the *marconista*. The wireless operator, unlike his telegraphic predecessor, has no time to interpret the series of dots and dashes inscribed by the Morse machine onto tape, as he is engaged first in hearing them through the headset and then, with a little practice, in instantaneously transcribing them onto paper. The wireless interface therefore couples a hand that writes with an ear that has been trained to capture acoustic data out of a noisy channel. The Marconi Wireless Company founded numerous academies across the United States to do precisely this: instructing potential listeners on what qualified as a signal and how to write it down. Unlike the distributed cognitive systems of today, which can run all by themselves without human interfaces, the wireless system of one hundred years ago required the mediation of the body-brain for the signals to be processed as communications.

Note as well that the wireless operator does not speak what he hears but instead utilizes a piece of paper that tallies the product of the wireless connection. Although Landini does not state so directly, the *marconista* makes sense not out of signs inscribed by the sheer force of the cabled signal as earlier telegraph operators had done but rather is called on to use his acoustic training to transcribe letters. He does not speak, or if he does, it is beside the point. What differs is the paper that registers the encounter between ear and hand. Much like Marconi, whom another biographer terms "neither a speaker nor a writer," the technical competence of the wireless operator consists in his designation of an economy of words in abbreviations, condensed phrases, and shorthand.[24] Such a connection of course requires a great deal from its human component, for to concentrate his hearing in such a way on a given signal so as to determine its tonality (and with it, its origin) meant solidifying the connection between ear and hand to such a degree that the *marconista* became "one with his apparatus." The resulting human-machine system sustains less of the opposition between the human and the technical because of the strength of their coupling. It may make sense to speak of overlapping fields of perception and registration that the wireless alone cannot produce.

Clearly this was not the case with the telegraph operator or the telephone relayers of the period since neither were required to take dictation of meaningful signals. The wired materiality of both media lessened the requirements made on its users, whereas radiotelegraphy at the dawn of transatlantic communication increased the demands on its user while finding its materiality in letters transcribed on pieces of paper. On the one hand, Landini seems to suggest that the Morse machine fails to discriminate "rationally" between atmospheric interference and very weak signals given its lack of the ear-hand coupling. On the other hand, the wireless incorporates the sense-making capacities of its human interface so as to increase the distance of the signal. It is both a device that transmits noise (the atmospheric static heard repeatedly) and a mode of exhausting information. In different words, both the telegraph and Marconi's wireless function as random generators of nonsense, "a strip muddled with indecipherable dots and dashes spit out by the Morse-machine," but only the wireless ensures the exhaustion of the transmission into sense. It restricts the

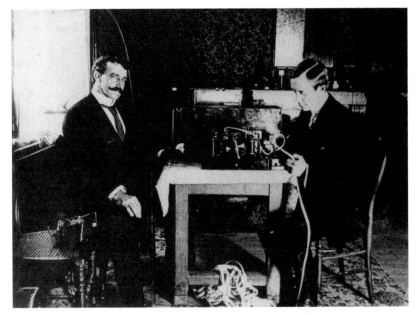

Marconi and the marconista *Kemp with an early Morse version of the wireless. Reprinted by permission.*

A late photograph of Marconi transcribing dots and dashes aboard the Elettra. *Reprinted by permission.*

media transposition of possible data to what can be heard as meaningful and written down, that is, to their symbolic transformation.[25]

Exhibits of Marconigrafia

The Monumental Marconi

With this preliminary investigation of primitive wireless practices complete, I want to turn to the cultural reception afforded the invention in a series of classic Marconi biographies. It will perhaps come as no surprise that the Marconi biography registers a terrific unease toward the invention and what it heralds for modern subjects. The principal apprehension concerns the mixed destinations of the wireless or, better, the consequences of the postal principle; anxiety about the wireless's alliance with modernity; and the status of the wireless interface, that hybrid creature, the *marconista*. The results are biographies that traffic in delirium until one is never certain who or what is operating which Marconi device, and a monumental rhetoric is deployed so as to allay precisely the fears of a wireless world. Thus, the Marconi biography offers mythic narratives of the sort Marshall McLuhan notes that rise up to meet all modern media: the literary conventions and rationalizations of narrative that are deployed to displace "a recording of the delirious effects produced by new media and infor-mation technologies" in John Johnston's formulation.[26] Principally these are the monumental status awarded Marconi the nonscientist and his device and the relation between Marconi and his telepathic, all-knowing mother, Anne Jameson.[27] Both counteract the onset of delirious destinations set in motion by the wireless.

The Marconi biography is particularly flustered by its inability to hear the wireless transmission properly. Since the biographies quickly transform Marconi into a wireless receiver and transmitter, Marconi's life becomes a hard-to-hear wireless communication transmitted by the Maestro himself, who listens in and corrects any erroneous details of his own life. The result is a series of dicta-tions between Marconi and his biographers that becomes the organizing center around which the Marconi legend grows. Landini, to take the first example, writes: "The episodes I am about to narrate apart from those unpublished, were confirmed and rectified by Marconi in the course of our interviews."[28] These writing sessions consist of responses shouted by Marconi to a hard-of-hearing Landini, who takes dictation; incredibly, all the sessions are offered as par-enthetical expressions during the course of their conversations. For his part, Dunlap announces his debt to Marconi more explicitly. Under the heading "Appreciation," we read: "For his [Marconi's] kindness in thoroughly reading the final proofs that the book would be accurate in facts about the wireless and historically correct in personal detail, the author is deeply indebted."[29] Here too Dunlap writes down copy but sends the final proofs to Marconi, belying his own self-description as author. Or better he forwards the book to the desire of Marconi with chapters such as "Birth of a Wizard," "As Triumphant as Caesar," and "The Hero of the Hour," in which Marconi is quoted in an odd, transmis-sive present. In the preface to his biography of Guglielmo Marconi, Luigi Solari promises his editor a book that will provide the reader with a "Marconi re-vealed" but quickly rescinds the offer. "But the principal object of this publica-tion of mine is not that of revealing Guglielmo Marconi, whose life does not offer much in the way of special revelations, a life prompted to great simplicity and great activity." Solari's account instead gives itself over to the monumen-tal status of the wireless and commemorates "the way in which this grandiose work was created and elaborated by him."[30] The narrative takes its cue from the impossibility of disclosing Marconi, lining him up on the side of the ineffable, and making him inhabit the silences the wireless creates.

> Out of the thirty-year history of the life of radiotelegraphy, what stands out in the tempestuous sea of electrical waves, radiating today from thousands of sta-tions, each one dominated by silence, that special spirit of international struggle, is the figure of Marconi emerging motionless and unbreakable like a rock.[31]

Solari's preface operates a transposing machine: on the other side of silence, a momentous spiritual struggle is being waged, out of which appears a petrified Marconi waiting for its ship to come in. Finally there is Degna Marconi's account of her father's life, perhaps the best example ever written of the tiny genre of biographies typed by the daughters of luminous inventors. The overall impression is the same: Marconi dictates messages to his *marconista* so that when the messages arrive, the contents have already been divulged. On the one hand, a feedback loop commences that allows Marconi to determine how well his secretary/listeners have heard him. On the other, a delirium of mixed and doubled destinations takes hold in the accounts that imitates the operations of the new communication medium.

The Mother of Wireless Invention

Sending messages to Marconi for rectification is only one way of ascertaining whether they have reached their mark. A far more complex system involves the assumed equivalence between wireless transmissions and the early dialogues between Marconi and his mother. The result is the remarkable pathos of rebinding these accounts continually foregrounded with regard to her and the monumentalization of the mother-son channel as utterly preceding the first wireless transmission. In the Marconigrafia imaginary, Anne Jameson, Irish singer, Italianophile, and relation to the Jameson distilling fortune becomes the explicit site for wireless transmissions *avant la lettre*.

These biographies offer numerous examples of Marconi's mother as wireless communication medium. Scanning back to a prewireless era when disappointing results were the norm, Landini imagines a young fretful Marconi asking his mother what he must do in order to transmit successfully. The encounter indexes similarities between the wireless transmission and mother-son communication, which return repeatedly here and elsewhere.

> "Come on Guglielmo, take heart," Anne said. "Why give up? You began so well. Don't you know that to live like a man means to struggle, to labour vigorously, tirelessly, especially when there's an ideal like the one you have to achieve. Always, my son, always move ahead, until you reach the end."
> "Oh, Mother."
> "Look around you Guglielmo, life itself is a duty, an exacting duty, [but] above all, it is a sacrifice; there is no place for the faint-hearted. Do you understand?"
> "Yes, Mother, yes."
> "So take up your work again, your tests. One day you will tell me if I was right or wrong."

Guglielmo, Anne, and Alfonso Marconi. Reprinted by permission.

"Thanks, Mother. Thank you [in English]. I was afraid that you . . ."

"What? What about [in English]?"

"Nothing, Mother. I was mistaken."

"Come on, take heart, take heart!"[32]

A number of details require scrutiny. First, one notes the persistent repetition in another language of the previous phrase, the frequencies tested to see if the signals have reached their destination. By all accounts, it appears the young mother continually alternated languages when speaking to Marconi, which quickly becomes proof of a mother sharpening his ears to pick out meaningful sounds. The fact that Marconi's first discursive repetitions were held in two languages destined for the ear of a bilingual mother takes on great importance as well; three if one counts the translation into a rhyming Bolognese that Landini, our "eyewitness," feels necessary to add parenthetically immediately after recounting the episode: "In describing the scene to me, the iron impassiveness of the Master was won over by emotion. I didn't know how to stop myself from exclaiming: 'Senator, you had a great mother.'"[33] These accounts would have Anne Jameson training him to hear the signal only he will hear.

Additionally, one notes the ease with which Landini, transcribing Marconi's words, makes Anne Jameson into a telepath. She, Marconi and Landini tell us, was a constant and faithful companion to her son, accompanying him to England when Italy refused to furnish funds for future tests, even taking dictation from her son when he filed his first English patent. The move from typist to mind reader is a short one for Landini: "But who better than a mother can read the soul of a son? And Marconi's mother was exceptional in this regard."[34] Thus, one finds Marconigrafia inscribing Marconi with signs that his mother reads and returns to him: once the wireless is up, the strength of these transmissions will be put to the test again and again, marking the power of Anne Jameson to read the message and her son's mind.

Although space limitations make a complete accounting of their relation difficult, what can be said here is that these accounts continually calibrate the proper distance given by the wireless to the mother-son connection, which they replicate in their descriptions of the many wireless experiments to come. Marconi's mother inhabits the distance that separates the transmitter from the receiver, which collapses at the moment of signal reception only to be distanced again in the next moment of transmission. These moments of connection and disconnection are analogous to that described by Freud and repeatedly cited in accounts of technological distantiation: "If there had been no railway to conquer distances, my child would never have left his native town and I should need no telephone to hear his voice; if travelling across the ocean by ship had not been introduced, my friend would not have embarked on his sea-voyage and I should not need a cable to relieve my anxiety about him."[35] The emergence of wireless communication strengthens virtual forms of interaction that become necessary when technical progress distances friends and family. As I soon indicate, virtual strengthening of familial ties is especially prevalent in accounts of the first tests of the Marconi wireless.[36]

Yet these chronicles of wireless invention differ from Freud's telephone to the degree they have the device contriving with a maternalizing superego that continually reconnects Marconi to his mother.[37] Not only do the biographies go to great lengths to characterize the reward Marconi receives on successfully registering the weak signal in maternal terms, but indeed we are repeatedly reminded that one of the first official tests of the Marconi wireless was to determine the proper distance between the Queen stationed on the Isle of Wight with the Prince of Wales, convalescing a safe distance from her on his yacht. The numerous attempts to increase the range of the wireless in the period between 1900 and 1902 has as its recompense the maternal voice in the Marconi biography, which Landini spells out in the remuneration implicit in the wireless transmission: "And the embrace of his mother . . . who had kept vigil and waited, feeling the same anxiety, the same trials, the 'bravo' of his mother must have seemed to him to be one of the best, if not the best of recompenses."[38] Throughout the descriptions of the various tests and the collapse and recalibration of distance, Marconi waits to receive the maternal "bravo."[39]

In the Marconi biography, however, Marconi's mother is never simply content to relay messages but also must put the young inventor in contact with Italy. Landini provides yet another example. Both Marconi and his mother are traveling to Calais and then Great Britain where Marconi has been summoned by the British Postal Ministry for further demonstrations of his invention.

> "Mother, what can I do for my country? You tell me," Guglielmo said with a disconsolate tone while the train wired *[filava]* its way to Calais. "Oh, my son, what are you thinking of? Yes, you will be able to do much for your country," Anne responded. "You can do so much with your device, with your example. Remember, my son [in English], that spiritual conquests and those related to work are worth more than military ones. Remember when I told you to be persistent and to move ahead, when I told you that one day you would be able to judge whether I was right or wrong? Well, today I will tell you the same thing, have confidence in yourself and move ahead. You will be able to do so much for your country, yes" [in English]. "Thank you, Mother [in English], I promise that I will do my best. I will do my best" [in English].[40]

The scene is striking for the connections it establishes between mothers and relays for messages emanating from the state. First, Marconi desires to hear from the mouth of his Irish-born mother what he can do for Italy in English and Italian. That Marconi should desire her voice to connect him to Italy is consistent within the accounts since it was undoubtedly a voice that carried Anne Jameson to Italy twenty years before: "Annie Jameson was a pretty girl from Ireland with a glorious singing voice and a will of her own. It was her voice that brought her to Italy to study *bel canto,* the trip offered as a consolation by her family because they had forbidden her to accept an engagement to sing at London's Covent Garden Opera House."[41] Marconi's desire for the signal is met by the voice of the other, which calls him, relayed through Anne Jameson.

Notwithstanding the difficulties of transmission, that is, the background noise reflected in the pondering repetitions, the passage draws a relation between listening to mother and listening for the signal, which metaleptically ushers in the presence of the state.

Of equal importance, the scene takes place in a train that "wires" its way to Calais. Another subterranean flash point in the history of the wired communications—the wireless and not the wires of late nineteenth-century Europe is enough to put Marconi on his way to Italy. What is it about transmitting without wires that in the Marconigrafia connects Marconi to the *patria* where other technologies failed? To answer the question requires both a short description of an epistemology associated with the wireless and the anxieties the wireless sets in motion. Landini provides an introduction from his childhood:

> But how beautiful it had to be, sending telegrams without wires. . . . And
> I looked contemptuously at the telegraph poles lined up along the side of
> the road, originating from who knows where and going who knows where.
> Sometimes I threw stones against the isolators, high above, and it so hap-
> pened that I smashed some of them really well, truthfully, I must confess,
> in tandem with other vandals.[42]

The episode is stunning in its reversals. Landini is uneasy about the presence of wires since he is uncertain where they come from and where they might be going, the *fort/da* of technological advancement made clear in the isolators that bring together signals and voices while keeping them apart. The contempt he registers is for the materiality of the wires, here marked by the poles that lurk above him. This suggests, it seems to me, that these accounts register anxieties rather different from those associated with wired technologies. Indeed the wireless itself becomes a vandal of sorts in Landini's account: by broadcasting it too seeks out the moment of disconnection in which its transmissions might substitute for the failures of wired communication. Of equal importance is that Landini, when registering the transmissive origins of wired materiality, fails to see the same operating principal at work in the wireless. Landini sees what can be seen, but does not see what cannot be seen: invisible waves are transmitted through the air without wires. This opens onto what Alice Yaeger Kaplan calls the invisibility of the wireless method that "made it subject to the greatest mystifications."[43] The first mystification: the wireless message arrives at its destination because its points of origin and arrival can be situated. Paradoxically, to hear even though there are no wires invests the wireless with an authority that contemporary radio listeners can only guess at. This, as I go on to argue in the following chapters, is no simple technological divestiture of the eye and ear of their authority, the one readers of McLuhan know so well, or simply a pecu-liarity of the Marconi biographies, but rather an effect of broadcasting without

Marconi, the wireless vandal. Courtesy of the Archives of the Fondazione Guglielmo Marconi, Italy.

wires that demands faith among its listeners. The Marconi biographies capture well the tautology: the wireless transmission is invested with an authority precisely because of its wireless (and not wired) materiality.

Wireless Paternity

How can one write about the wireless without compounding its destinations? There is the feedback loop in which the Marconi biographer is always sending a marconigram and the monumentalization of a maternal transmission as preceding the moment of invention. Reading these accounts of wireless invention, however, the answer may be as simple as devising narratives that cover doubling operations so as to deflect curiosity surrounding the device. Landini does so by associating the wireless with inspired artistic productions. The wireless is "a discovery, an invention, a literary, artistic and scientific creation" that always contains "a little something of the Divine."[44] For his part, Dunlap picks up "the long threads from which the odyssey of Marconi is spun," weaving a web of attunement between Marconi and the Creator's grand design. "Genius had to discover in the Creator's plan a mysterious medium hidden until the twentieth century approached, when mankind's accelerated pace needed communication on a worldwide scale; more rapid than the mail carried by an ocean liner, faster than an airplane dispatch, quicker than the telegraph of Morse or the telephone of Bell."[45] The wireless superman requires no introduction, but consider rather the ease with which Dunlap construes speed, one of modern life's primary characteristics, with a transcending "Creator's plan." One master narrative is joined to another, industrial capitalism, so as to account for the appearance of the wireless. Dunlap weaves his web, relating technological advances to each other and linking the same transcendental plan with technology's advances.

A second thematic revolves on the Hegelian/Darwinian axis: "God gave intelligence to man to evolve and not to stagnate," writes Landini, "so that man might give himself to discovering the mysteries of nature utilizing his courage."[46] Thus, Marconi's compulsion is one with the essence of the wireless itself: "The case of her son was not that of a run-of-the-mill man . . . He always had to be on the march, incessantly ascending: it was his inclination to do so, just as in his spirit and in the invention itself."[47] Since Landini and Dunlap regularly forego the distinction between operator and wireless, they must instead find another form with which to observe early radiotelegraphy. Thus, they elaborate the historical difference between greater and lesser alterations of humanity. Scenarios based on wireless gains for humanity alter history, tilting the scale to Marconi and Edison: history becomes a series of technologically derived epochs that remake humanity. So Landini concludes that "humanity is not transformed as much by Alexander the Great, Caesar or Napoleon, so much as by Volta, Edison, Faraday and Marconi."[48]

The difference in significant transformations of humanity that marks monumental and technological histories requires that the latter find time to carry out evolutionary schemes. Consequently, the biographies redound with calls for continued human evolution to rise to the challenges presented by the wireless. Dunlap argues, for instance, that Marconi, Bell, and Edison "pioneered in the nineteenth century—in the Victorian era, when rugged individualism was a potent force in science."[49] Both biographies attempt to persuade readers that technological advances arise out of human achievements, simultaneously desiring and fearing a transhistorical phenomenon connected to no one moment or invention.

One can also deflect curiosity by simply insisting on Marconi's sole paternity in the matter. The wireless does not just electrify Marconi, but indeed Marconi *fathers* the wireless. Dunlap contrasts Marconi's paternity of the wireless with the promiscuous state of discovery in 1937: "Today the glory of discovery is divided. No one man wears the crown. . . . Science now scatters and shares the inventive laurels; it emblazons the name of no lone inventor."[50] Yet throughout the account, Dunlap often cites Marconi's debt to earlier inventors, as does Marconi himself. "I am uncertain as to the final results of my system. My discovery was not the result of long hours and logical thought, but of experiments with machines invented by other men to which I applied certain improvements."[51] Landini also sings the Marconi paternity blues, recalling how Alexander Popov, who had experimented with Hertzian oscillations before Marconi, was forced to recognize Marconi as the father of radio. One also notes the frequency with which both biographers name the wireless a child of Marconi's, a "spark-shooting embryo" in Dunlap's words, which links it to a chain of existence and technology associated with the lone inventor.

And if the wireless should fall to divided discovery, to use Dunlap's terminology? Fixing the wireless to Marconi serves a number of purposes. It lessens the uncanniness that surrounds its emergence, helping to obscure the lack of wires that connect Marconi, Popov, and Tesla to the wireless. Moreover, even a cursory reading of the accounts reveals not a little technophobia lurking at the margins. The fear is that the wireless is a new form of machine that is outside of history, with no wires connecting it to one inventor or to one moment. Consequently the wireless is figured as a technology out of control and at the controls, threatening to "divorce Marconi's name from the 'ever-expanding' invention."[52] And thus the repeated attempts to rejoin Marconi's name to an invention and the failure of narratives of progress and paternity in containing the uncanniness of the wireless; proof that the translation of a transmission medium, the wireless, into another, writing, often exceeds the possibilities for narrating it.

Acts in an Occult Drama

Transforming Marconi into a monument or superman or turning Anne Jameson into a wireless transmitter *avant la lettre* are not the only modes by which these biographies register and elide the effects of the wireless system. When that strategy is no longer available to counteract the ease with which the wireless exceeds its narrative bounds, Marconi the magician emerges. There are, of course, the early hauntings Marconi performed in the villa before the wireless fully comes into its own; invisible hands frequently shadow the wireless by moving objects, alarming the neighbors unnecessarily. Indeed Landini recounts that many of Marconi's childhood friends believed the house to be haunted since bells rang at all hours for no apparent reason, leading to the outbreak of a "telegraphic epidemic" of fear across the periphery of Bologna on news of the first successful transmissions.[53] Given the biographies' penchant for transforming Marconi and his invention into exhibits in turn-of-the-century magic, it is not surprising to find that the wireless was born wireless, the exact date of its temporal inception missing from the official records. Dunlap cites Alfonso, Marconi's brother, to the effect that Guglielmo actually made hundreds of tests before moving the device out of the laboratory, including repeatedly ringing the front door of the family's villa from his third-floor room.[54]

A qualitative leap in production values occurs once Marconi takes his wireless to England. There his black bag and top hat make him a double for either a traveling magician or terrorist: "Incited by Guglielmo's replies to their questions, which, as they were not understood seemed evasive, and their own dutiful sense of importance, they manhandled the foreigner's mysterious box. Its contents were broken beyond repair."[55] No backdrop was more remote-controlled, however, than that of Stonehenge:

> A first step was taken in that direction when Marconi's modern machinery, his kites and poles, went up in the spiritual shadow of ancient Stonehenge. On the rolling Salisbury Plain, the sending apparatus was housed in a shed they called the Bungalow, and the receiving apparatus was placed on a hand-drawn military cart to give it mobility. Roughly made copper parabolic reflectors were mounted on each side of the detector. The first signals came in from a distance of a hundred yards, the next covered a mile and a quarter, then six and finally nine miles.[56]

In the shadow of the Druid ruins, the biographies multiply the uncanniness of the wireless when it takes its first steps, hidden beneath the first mobile land receiver. Given the stage, Marconi is quickly transformed into a renowned magician. In Dunlap's aptly titled "Acts in an Occult Drama," Marconi is "a practical-minded scientist," who in an elongated dash crosses genres and audience to become "a master magician who could juggle coils of wire, unleash electrical impulses and pull messages from the air, all in the twinkle of an eye." The condensers, coherers, antenna, and *marconista* were but "props assembled before the footlights of the scientific world."[57] Staging spectacles before the footlights of science and the British Postal Service is easily conjoined to the possibility that Marconi is a witch and his craft witchery: "How could mute devices built by man cause the immensity of space to pulse with living sound? . . . To do it would be witchery," asks and answers Dunlap.[58] Marconi also recognizes that the wireless assumed a certain staying power from a struggle to keep down the spirits of an inquisitorial past, often repeating his thanks that he had not been born during the Middle Ages for surely he would have been burned alive.

Such anecdotes point to a genealogy of the wireless within the shifting frontiers of art and the occult. The lines running between Marconi and Frankenstein are a case in point. In attendance at the Stonehenge transmissions and throughout most of the wireless trials in England was Sir William H. Preece, chief engineer of the British government telephone service in the 1890s and a crucial figure in the development of the wireless in Great Britain. Before reaching knighthood, William Preece was also a young apprentice to Andrew Crosse, an electrician made famous for his attempts to create insect life by passing electricity through stones. Many believed Crosse to be the inspiration for Mary Shelley's *Frankenstein*.[59] Thus what Marconi's Stonehenge performance shares with magic is precisely its staging "of a miraculous thing that contorts, condenses, or somehow usurps partial powers of immortality in profound complicity with the supernatural."[60] A systematic demonstration is beyond my present resources, though describing the relation between the wireless and Nature is useful, particularly when the concern shifts to miraculous accounts of wireless impregnation.

Seminating Signals from Radio Marconi

Given the wireless's capacity to penetrate all obstacles, the Marconigrafia's immaculate wireless conception will surprise no one.

> A woman who lives in a little English town has been writing Marconi for years, every day without ever having met him. Marconi has never seen this woman and has never responded even with one line. . . . But one fine day Marconi received a photograph of a beautiful baby, and these words: "Here's your spiritual son." It was then that Marconi began to worry considerably. He told the English police, who after a thorough investigation reported that the woman in question was even-tempered and honest. She declared she was in love with Marconi without knowing him and that she admired his photograph every night.[61]

The anecdote and others like it claim for Marconi and his wireless, a capacity for semination well beyond its years, again associating the ear as a privileged organ of reception and its impregnation by the wireless. Not only is scattering wireless seeds the flip side of Marconi absented by his device; Marconi is sufficiently present in the transmission to generate thought and semen. In different words, the Marconi wireless is present enough to inseminate artificially but absent enough to have a reasonable alibi. This performative aspect of the Marconi wireless challenges Nature throughout the Marconigrafia by providing Marconi with the occult powers of creating life via ear impregnation. The biographies read as glosses of Derrida's postcard that impregnates since the wireless does not require checks drawn on some sperm bank but "remains living enough for the artificial insemination to yield fecundation, and even desire."[62] The happy choice of "broadcasting" among early wireless users to complete a radio lexicon confirms the overdetermined relation between insemination and the wireless.[63]

It also delivers miscarriages by means of cut messages.

> At 3:15 in the morning, we clearly received the following message from the English station at Poldhu: "Espress (*sic*) of Russia had a miscarriage." "Bravo Marconi," exclaimed Mirabello. "Your creation is now complete." Finding myself two years later in London at an international radio-telegraph conference, I was invited to lunch by the delegates of some foreign governments. . . . "It's strange," I said, "that the news of such a great failure served to confirm such a great success." "Hear, hear! You are a jolly good fellow," the English shouted.[64]

The extent to which the Marconi biographies figure the wireless as an explicit seminating force or a technological Moloch to whom human sacrifices are required is remarkable. Missing letters take on the aspect of flesh and blood; a missing "m" in "espress" is echoed in her failure to give birth, the messenger having been exchanged for the message. More important, Marconigrafia will find a motto for the success of the wireless in the formula that great failures confirm great successes. Again and again the wireless demands miscarriages in these accounts but prefers larger catastrophes, for example, the sinking of ships and states, to demonstrate how well it picks up the signal. The wirelessed miscarriage of the Russian empress looks forward therefore to greater catastrophes to come.

Mourning over the Wireless

Earlier I described the function of a mother's voice in accounts of wireless invention and with it the maternalizing force that underlies many of the experiments.

In this section, I want to return briefly to that theme in order to broach the question of mourning implicit in idealizing a mother's image. Such a discussion turns on the biographies' positing of Marconi and his daughter's bodies as theaters in which wireless processes are registered. The wireless, much like the telephone and telegraph before it, maintains a disconnection between its points of transmission while propping up an acoustic memory that is meant to maintain the link to what has been lost. Interestingly, the biographies translate the loss and separation as a mode of object preservation. Vaguely sensing a genealogy of the wireless in the technology of meat preservation, Dunlap accordingly recounts the story of a Napoleonic-era butcher "too fat to fight but keenly alert to the tortures of hunger," who chops up beef, pork, and veal and stuffs them into the intestinal tubing of a pig.

That was not all.

> He made his invention practical for soldiers on the march. He linked the sausage together so that long strings could be hung from the neck or wound round the waist. He tied this portable food supply into short lengths for convenient ration. As a popular idea it swept the world; appropriately, to this day it is called bologna.[65]

The short links, for convenience's sake, mirrored textually in the repeated "he made," "he linked," and "he tied," disclose a particular orality that runs throughout most Marconigrafia. The wireless and bologna assume one cannot take it all in and so preserve the beloved in morsels spread out over time. The wireless, inasmuch as it carries the idealized image of mother, preserves. And as preservatives make swallowing losses difficult, it operates along the same lines as painful orality, which accounts for the frequency with which mouths are coded with pain in each of the biographies. Marconi guarded his mouth jealously, constantly limiting his speech and the food that entered, which D'Annunzio captured ironically when he doubled Marconi for Arpocrate who seals lips with a finger.[66] Solari's biography of Marconi registers a similar distress at the dawn of wireless transmissions:

> The transmission of the letter "S" was like a pinprick in that central nervous system and made all organisms based on the functioning of cables cry out like eagles. And the scream let out by these organisms directed the world's attention to Marconi's invention.[67]

The pinprick that makes existing life-forms cry out becomes the means by which the wireless reproduces itself across the body of the wireless operator: the central nervous system of wired organisms (read *marconista*) is punctured in a moment that precedes the connections between device and organism that form the wireless interface.

Marconi's enormous interest in his daughter's mouth suggests an oral fixation, as do his attempts to force open her hermetically sealed lips to help the medicine go down: "All else failing, Mother and Father resorted to the cure prescribed by Lauro—champagne. Father had to force it between my lips, but not for long. After one gulp I began to revive and took my medicine with avidity."[68] So too do her teeth occupy him to no end: "There were the usual school-girl problems, of course, and Father wrote Mother in October: '. . . As you know, while in Paris I went twice to Dr. Hipwell (once with Degna) about Degna's teeth.' . . . In November the problem of my teeth still loomed large."[69] The mouth's traditional zoning patterns are still recognized (it receives food and remains the site of language), but now it is the site of a rezoning: Marconi's incursions into his daughter's mouth penetrate the space that separates the body's inside from its outside, but they are unilateral. At first glance, Marconi's anxiety might appear to cover the same ground as that traversed by the telephone before it but with a significant difference: Marconi is far removed from Bell's speaking exercises for the deaf, which as Avital Ronell points out, trained mouths to speak and, following Derrida, to listen.[70] Rather, mouths in the early wireless network receive and only infrequently transmit. The history of the wireless operators confirms this since their function was to connect hearing to hands that recorded symbolically the dots and dashes that scurried across the ether; only rarely do they initiate discourse. The substitution of mouths for ears is grounded in early wireless practices and not simply as the doubling of ears and mouths as sites of penetration but opening up a space from which new forms of subjectivity (and their prerequisite bodies) can emerge.

Not surprisingly, Marconi biographies register their unease with wireless interfaces by forcing these functionaries of the new communications assemblage to maintain radio silence, keeping their mouths shut since nothing is heard until the mouth closes around the wireless bit. The contrary also holds for Dunlap, Landini, et al.: in order to transmit, mouths must be closed. The transmission that results is one based on a nonspoken, nonpulmonic, and nonhermeneutic system. The association is not frivolous as a quick glance at Upton Sinclair's unjustifiably ignored *Mental Radio* demonstrates.[71] There the wireless allies itself with mental telepathy in an imaginary mapping of the changes bodies undergo when wireless devices are simulated in an author and his clairvoyant wife: whenever perceptual capacities are extended, the hope (or threat) of simultaneous mind reading is never far behind.[72] A similar move occurs in the monumentalization of Marconi's own body in these narratives: not only is the wireless made to seal Marconi and his daughter's lips, but it also requires acute hearing. This perhaps accounts for the extraordinary relationship between Marconi's ears and his device in the accounts, the unconscious registration of the modified relation between the wireless and the ear I mentioned above.

"'Che orecchi grandi ha!' (What big ears he has!) exclaimed an old servant of the house on seeing the little stranger for the first time. And what historic sounds they would hear when science tapped on their drums."[73] Indeed, after long years of practice in listening for wireless dots and dashes, the long waves of science, and later the shortwaves of fascism, Marconi had "sharpened his ears to an unbelievable extent. Uncanny is his facility in hearing conversation at a distance."[74] Underscoring the peculiarities of the human component in the wireless interface, Dunlap tells us that Marconi "seems to have a multiplicity of eyes, yet he has only one" (the other lost in a bizarre motoring accident). Thus he can search "the skies for information, for new scientific clues."[75] Denial is built into the argument since Marconi possesses superior vision although he possesses only one eye, much like Bell who inherited his keen hearing from his half-deaf mother. I will have more to say about Marconi's ability to see the future when we turn to the speech D'Annunzio dedicates to him at Fiume.

Extending his observations to Marconi's head, Dunlap uncovers a remarkable resemblance to the antenna.

> So long as he keeps his hat on he is quite like other men. But when the hat comes off Marconi's personality manifests itself. The real strength of the man is revealed in the peculiar shape of his head. . . . Behind the ears it is a flat head—the head of [a] shrewd man who eats to live not one who lives to eat.[76]

The study of the relation between craniums and intelligence recalls the work of Cesare Lombroso and his studies of degeneration. In this regard, we turn opposite the title page and there, equipped with a skipper's cap and enormous ears is Marconi and the ironic caption: "Who says Italia holds a dying race / And all the glory of her line is spent?" The technological makeover of Marconi, the first wireless operator, does not begin and end with his head, but also extends to his tapping finger. Thus the biographies detail Marconi's remarkably long and agile fingers, tested for hours at the piano. One in particular deserves attention: the finger that taps the wireless key. Was it the same finger that Marconi detached and reattached as a child with such joy, as his cousin remembers to Degna Marconi?

> When he was sixteen he was so fascinated that he built a machine of his own for transmitting electricity. Very rough it was, too. Daisy Prescott, who came in while he was constructing it, found him using blunt scissors to cut wire into half-inch pieces. . . . Not surprisingly he managed to cut not only the wire but a piece off his finger with the "treacherous" scissors. He stowed his sacrifice to science in a box and remarked airily, "I'll get it stuck back on by and by. The chemist's not far from here."[77]

The passage is worthy of Mary Shelley: science demands that sacrifices be made, that fingers be cut, and offers the means to reattach the missing finger. In Marconigrafia, the wireless reenacts the sequence, having a finger materialize some distance away. The initial cut allows it to double for another finger that transmits the "S"; when detached from the hand, the finger successfully extends the reach of the transmitting wireless. The Marconi biography dissects the wireless and finds an eleventh finger or piece of finger.

Marconi, *Marconista*

This chapter, dedicated to the birth of radiotelegraphy, explored both the material practices associated with the dictation of meaningful signs that are heard and written down in an emerging wireless media ecology and the delirious effects engendered by early wireless technology. In the first instance, the classic experiments anticipate what will be required of the *marconista*: the acoustic reception of weak signals, the solid connections between ear and hand required to transcribe the signal, and his subsequent inscription as a component of a wireless communication network. We should note in particular that the primitive wireless circuits detailed here transmit acoustical streams of data alone and not the optical or the previously written: only that which can be distinguished by its being written down (and not simply copied) will correspond to the radiotelegraphic threshold for recording. Put otherwise, the wireless network defines its own information processing as the pattern recognition of minimal signifiers that a *marconista* registers.

In the second instance, Landini and company use the conventions of the traditional biography to transpose these troubling features of wireless communication generally. Not only does the wireless unsettle notions of who is properly sending and receiving a message—a worrisome aspect that leads Marconi to begin prophylactically contacting his biographers—but it often exceeds the narrative protocols of the biography itself. The Marconi biographers read into Marconi's past a strong sensibility to disconnection and therefore either obsessively foreground the monumental features of the device or emphasize its enmeshment in discourses of the occult and magic. The prodigious features of Marconi, the boy inventor, his refashioned wireless body, and Anne Jameson as wireless medium: all are imaginative strategies that imply the power of wireless technology at the turn of the last century to shake up categories of communication, writing, and distance. What's more, the biographies deplore in their own way the intensity of changes set in motion by the wireless and dramatize the process whereby the *marconista* comes into being. The perceptual changes that result with inscription into a wireless network become the shadow moments to which they respond. In their sealed lips, telepathic moments, and wireless semi-nation, Marconigrafia are implicitly marking the stronger human-machine coupling posed by the wireless invention while creating narrative spaces in which the delirious possibilities the wireless sets in motion are neutralized. The result are biographies that recode alterations in perception as delirious mani-festations of doubled identities, the principal being Marconi for the *marconista*.

What happens when our perspective shifts to mobilizations of the wireless in narrative arenas that share little of the rationalizations and monumental pro-tocols of the biography? D'Annunzio's speeches at Fiume are a historic case in point for there the poet mobilizes the desire and the fantasy of the legionnaires by recoding the occupation as a wartime marconigram sent by D'Annunzio to himself. The gripping result is a series of speeches that deploy the wireless to create an apocalyptic tone that promises to reveal the contents of the Fiume marconigram. It is to that wireless message that I now turn.

2

D'ANNUNZIO AND THE "MARCONIGRAM": CROWD CONTROL AT FIUME

A T FIRST GLANCE, a chapter dedicated to Italian poet, national bard, and prototypical dandy Gabriele D'Annunzio may seem far removed from an itinerary marking correspondences between early wireless technology and cultural production from the same period. The dictation that *marconisti* were required to take when interacting with the sounds emanating from their headsets would have been strange for a poet who could never overcome "my old aversion to dictating" or overlook the "secret modesty of an art that rejects intermediaries and witnesses which stand between the subject and him who works with it."[1] Moreover, in its economy of transmission, the wireless neutralizes the discursive effects of words and in doing so generates a wireless operator whose acute sense of hearing obscures his failure to read the dictation he has taken down. D'Annunzio's prose from *Il piacere* to *Il fuoco* to *Forse che sì forse che no* rarely practices an economy of words: in fact, quite the opposite, as Renato Barilli has argued.[2] Thus, most accounts of D'Annunzio's literary production forego any medialogical context, focusing obsessively, in the case of

Italian criticism, on his "supposed incoherence, capricious discontinuity, and temperamental volatility," all of which make him perhaps the most repetitive writer in Italian literature.[3]

Thus D'Annunzio would appear to be of peripheral interest in the age of Marconi were it not for "l'esperienza fiumana," the crucial years between 1919 and 1921 when he, along with large numbers of recently demobilized (and not so demobilized) war veterans, occupied the former Austro-Hungarian port of Fiume. Hoping to alter the victory "mutilated" by the Versailles Peace Treaty through a territorial fait accompli, D'Annunzio and his legionnaires, as he called them, vowed to remain there until the Italian government enforced the concessions won at the Treaty of London that had enabled Italy's entrance into the war; one of which being precisely territorial expansion in the Balkans. The longer the occupation lasted, the more D'Annunzio's self-proclaimed adventure in diplomacy was transformed into an experiment in political reform and mass politics. A constitution was drawn up that guaranteed women the right to vote while D'Annunzio and his council of ministers promulgated anarcho-syndicalist reforms and guaranteed artistic freedom of speech. That the occupation ultimately failed to change the terms of the peace is not unexpected given D'Annunzio's failure to win over the Italian state or the larger Italian public. When Italian naval vessels shelled D'Annunzio's headquarters on January 18, 1921, the Fiume *comandante* (as he was officially known) quickly fled the port as did most of his followers, effectively ending the port's occupation.[4]

Despite or perhaps because of its ultimate failure, Fiume continues to generate critical heat, especially among those interested in genealogies of Italian fascism. For many what weighs most is D'Annunzio's conception of politics, which "was his major contribution to fascism in Italy," and here Fiume seemingly confirms their views.[5] The emphasis thus falls on the political drama of the Fiume experience and, in particular, on the secular rituals that accompanied D'Annunzio's numerous speeches to the masses. Speaking from the balcony that rose above the crowds and surrounded by blood-red columns, flags, and the symbols of Italian nationalism, D'Annunzio becomes for Michael Ledeen a sort of operatic dictator: "In this sense, Fiume was not only a precursor of fascism's extraordinarily successful enchantment and manipulation of the masses. By infusing religious ritual and symbolism into political events, it was a model for much of the mass politics of the last century."[6] Consequently, Fiume operates as a microcosm of the "madness and the magic of the twentieth century, an early laboratory in which the germs of mass politics—of both right and left—were tested on human subjects."[7] The correspondences with fascism—though more precisely with Mussolini—consist principally in both D'Annunzio's and Mussolini's ability to convince their followers to subject themselves to *il duce*, whether it be through hypnotism, magic (again the choice of words is Ledeen's), or through a contagion of mythic perception that shuts off the subject's critical faculties.

What is so striking about attempts to think about Fiume and Italian fascism is the marginalized status repeatedly awarded the speeches that D'Annunzio gave in the months leading up to Fiume and those directed to the crowds throughout the occupation itself. Attention continues to hover over the psychological climate that made Fiume the "City of Life," "a sort of experimental counterculture with ideas and values not really in line with then current moral codes," or to the sheer magnitude of the enterprise, "the first example of subversive adventure nourished by the aftermath of the war."[8] Missing is an engagement with the speeches and their success in creating conditions that allowed D'Annunzio to become "the interpreter and accomplisher of *fiumanismo* that otherwise would not have gone much further than the disordered phase of confused mental states."[9] It is here that an analysis of the media ecology of the period may prove useful in underscoring how the speeches enact a form of technological hearing analogous to wireless practices described in chapter 1. Put differently, we will want to ask how the speeches, confirmed in the numerous testimonies that chronicle the eighteen-month occupation, register the operations of constructing subjects at Fiume and what similarities among the crowds at Fiume, the *marconista,* and the future fascist are the most significant.

To answer these questions requires that we ask who precisely is listening to D'Annunzio at Fiume, what narratives the speeches assemble that could give rise to new acoustic worlds and the subjects that inhabit them, and how we might circumscribe "the message of Fiume" as the product of a media ecology altered by the introduction of the wireless. Accordingly, I describe the crowds that gathered at Fiume and detail some of the reading habits they had acquired during the war that might account for their ability to take dictation so ably near the Palazzo del Comando, wartime alphabetization campaigns being of particular interest. I then examine how the speeches inscribe Fiume into a narrative of defeat and victory so that the city is transformed into "la città olocausta." Most readings of the speeches find D'Annunzio deploying a Christian cosmology in the choice of symbols, the repeated references to Pentecost and the chalice that could not be refused. "What is striking about the Fiuman writings," for Barbara Spackman, "is that they are dominated by a Christological rhetoric. Nothing of this kind appears in his novels and poetry where Catholicism is of the decadent eroticized sort and Christian lore is treated with a sneer."[10] Agreeing with Spackman's insightful reading, my emphasis falls on the apocalyptic tone that the speeches evince. I show how D'Annunzio translates the message, Fiume, as one of sacrifice and salvation in which those listening are required to wait for the message and then to act. Buttressing my argument with examples from D'Annunzio's *Notturno,* a memoir written immediately before the Fiume occupation while

D'Annunzio was convalescing (but published afterward), I delineate the borders separating sensory deprivation and technologically authorized glimpses of an unknowable future.

As for associating the message of Fiume with an altered media ecology, once again the figure of Guglielmo Marconi precedes us, for Marconi was present at Fiume in the waning months of the occupation. Steaming into port on September 22, 1920, aboard the aptly named *Elettra*, Marconi lent his "moral" weight to the entire enterprise by appearing with D'Annunzio on the balustrade, providing star power for the speech D'Annunzio gave in his honor, and transmitting a wireless message from D'Annunzio asking that Fiume "remain joined to the mother Patria *in perpetuem*."[11] D'Annunzio's speech, utterly ignored in commentaries on the significance of Fiume, becomes the key by which I decode the Fiume message as an apocalyptic marconigram that reinscribes hearing within another code of meaning, that of sacrifice and death. Indeed D'Annunzio deploys the wireless and Marconi in ways not so different from Dunlap and company: both authorize access to hearing, but where the former merely emphasize the occult and magic in their delirious biographies, the wireless at Fiume allows D'Annunzio and the crowds to think—if that is the term—a deliriously hyperbolic future for Fiume. D'Annunzio mobilizes the wireless imaginatively in order to sanction his status as privileged relayer, modern prophet, and *comandante*. A reading attuned to the medial qualities of the speeches will find the wireless providing cover for D'Annunzio's prophetic claims while also contributing a model for the kinds of deterritorialized perception that were so prominent in the creation of the *marconista* of the last chapter.

The notion of a message that carries with it the possibility of programming listeners via an apocalyptic tone becomes in the chapter's final section the fulcrum by which I set forth possible hidden topographies of fascism and Fiume. Of course many have noted how indebted fascism is to the Fiume experience. Generally, attention centers either on the balcony address, the shouts of "aia, aia, alala," the dramatic dialogues with the crowd that they share, or their coterminous beginnings in the ashes of the First World War. In Emilio Gentile's pithy analysis, both partook of the war that "bore new myths of a national community: the ideal of comradeship, the worship of the fallen and the war heroes, the exaltation of individual courage and collective discipline, the apology of a new hierarchy based on personal qualities of leadership."[12] Highlighting the characteristics of the marconigram first transmitted at Fiume, I hope to establish continuities between Fiume and a fascism that knew well the advantages of generating alarm in its subjects. Some may argue that such an analysis fails to respect the cultural and political differences that inhere between an anarcho-utopian movement that lasted all of a year and a half and a regime that remained in power by force of arms for twenty years. Reducing both to an apocalyptic

D'Annunzio broadcasting his wireless appeal from Fiume. Reprinted by permission.

register generated by a form of wireless transmission smacks of technological determinism. A response, it seems to me, would have to show how both Fiume and fascism mobilize desire and fantasy in such a way as to create conditions in which something like an attraction to sacrifice itself is produced, in a message that switches off critical faculties.[13] Rather than technological determinism, the suggestion is that the power of sacrifice and death among fascist adherents is in no small way related to the medium by which it is transmitted and the attendant apocalyptic tone that is generated.

Narrative at Fiume

The Arditi, *Indoctrination, and World War I*

The first step in determining how D'Annunzio's speeches simulate acoustic moments that reinscribe hearing is to examine more closely who constituted D'Annunzio's public at Fiume. Not surprisingly, given the sheer numbers of soldiers under arms in Italy during the eighteen-month occupation, the vast majority of D'Annunzio's Fiume followers came from among their ranks:

> By the summer of 1919, there were 1,575,000 men still under arms (excluding officers) in the Italian Army. Slightly more than half of these men were under

the supreme command, and of these, some 740,000 were in the war zone, deployed against the risk of new hostilities—a distinct possibility at a time when the Italian delegation had temporally withdrawn from the peace talks in Paris.[14]

Of these numbers, historians paint a picture of an army in disarray: "battalions in revolt against their officers, generals arrested and insulted, commanders who confessed that they did not know whether their men would obey orders, soldiers who shot at other soldiers."[15] D'Annunzio drew most of his legionnaires from among these men, so many in fact—at the height of the occupation he had nine thousand—that he had to appeal to Italian soldiers to stop defecting. In addition, a significant portion of D'Annunzio's adherents came from the ranks of the *arditi,* or what today we might call special forces, as D'Annunzio himself noted after the occupation: "When they speak to me of misunderstandings between the *arditi* and the legionnaires, I revolt at such an absurdity. Weren't my legionnaires composed of three-quarters *arditi?*"[16] The origins of the *arditi* as a fighting unit remain clouded. Introduced in the last part of the war, these troops "conscripted on the basis of a particular selection process, subjected to the severest of trainings, equipped with a special armament adapted to rapid incursions and hit-and-run assaults," enjoyed a reputation for invincibility and extreme ferocity in battle.[17] Their newness consisted "in the strong motivation for battle that distinguished the members of these divisions from the large majority of soldiers," though other characteristics soon became apparent.[18] In its concluding pages, F. T. Marinetti's war memoir *L'alcova d'acciaio* chronicles a number of assaults conducted by the *arditi:* of particular interest are their automatic responses, suggesting a kind of mechanized and preprogrammed behavior.[19]

As interesting as the history of the *arditi* and the trench experience of regular Italian grunts is, what stands out is the urgency with which the Italian state set about constructing a battle-ready army in the space of three years; one year if we limit ourselves to reforms undertaken after the disaster at Caporetto in 1917. One of the most significant changes concerned its campaign to create literate subjects whose sense of nationalism could be nourished by the introduction of trench newspapers, the printing of propaganda, circulars, and of course books. The success of the campaign was astounding: "In 1921 in only two regions (as opposed to the seven of 1911) did illiteracy remain above 50%, while those in which it fell to below 13% were now five (with only Piedmont enjoying that status in 1911). On a national level, the rates of illiteracy fell in the same period from 48.5% to 27.4%."[20] The war forced a population that before was profoundly oral-based to alphabetize itself in short order. The access to written forms of communication and the writing and reading habits that developed signified a monumental change in social communication, most importantly in the ability to address communication to a host of contemporaneous destinations. The outbreak of politicization among troops at the end of the war may in fact have brought on greater rates of literacy and the rapid diffusion of cheap printed matter.[21]

Much of D'Annunzio's repeated calls in the speeches to a unified populace at Fiume are conditioned by the Italian state's success in utilizing alphabetization to create a large mass of motivated men. By addressing itself to soldiers in the form of pedagogical tracts and primers, the state had certainly "embarked upon a campaign of indoctrination, convincing the soldiers that they had a deep personal stake in the outcome of the war."[22] Of equal interest, however, are the effects of the campaign on writing, and in particular on the obsession with taking dictation at Fiume. Eye-witness accounts of the occupation are filled not simply with listeners blinded by D'Annunzio's rhetoric but also with soldiers taking dictation of the speeches and readers of D'Annunzio's prose imitating the poet's style in their own poetry.[23] Admittedly, it is impossible to determine with any degree of exactitude just how widespread the phenomenon of mass stenography was at Fiume or, for that matter, what precisely the legionnaires were reading aside from the newspapers and civic proclamations published by the occupying forces. Still, crowds taking dictation transform the context and circumstances of their interaction with the poet's words since meanings are transformed when listeners also emerge as readers, writers, and copiers.

The novelty of the kind of interaction between the literate and semiliterate crowds and D'Annunzio should not be overlooked. First, a feedback loop emerges in which the *comandante* could judge how well his words had reached their destination, what in the oration "Santa Barbara" becomes the power "in every instant, to convert our cry into action, to convert the word into blood."[24] An expanded data pool regarding what appeals were more successful enabled D'Annunzio to tailor his speeches in order to generate greater attention. Indeed, reading D'Annunzio's speeches chronologically from the "Lettera ai dalmati" to the last appeal to the gathered crowds (*fiumani* in the original) is an exercise in how attuned D'Annunzio was to a rhetoric capable of generating action. George L. Mosse suggests as much in his elaboration of the relation between D'Annunzio and the legionnaires:

> At one point when D'Annunzio had given a speech to his legionnaires, young soldiers and officers spontaneously produced their own confessions of faith couched in the poet's lyrical style. Here, on another level, D'Annunzio's prose itself provided the link with his audience, the *prose d'annunziano,* with its exaggerated rhythms and pomposities, dominating the consciousness of hard-boiled and semiliterate legionnaires.[25]

Mosse intimates that the means by which "consciousness" is dominated includes the written, though unfortunately his analysis trails off at this point. The implicit link between D'Annunzio and the legionnaires, however, is one of mimicry and the creative act, which are typical of some of the effects of industrialized printing and alphabetization.[26] In delimiting the relation between D'Annunzio and the legionnaires as one of technologically authorized prophecy and apocalyptic tones, we ought to take account of the complicated links that join the two in their choral back-and-forth. This is not to ignore, to be sure, the severely divergent forces of left and right that battled for control at Fiume over the course of the occupation or the divisions in the ranks of the legionnaires between those *ragionevoli* (reasonable) and the *scalmanati* (hotheaded).[27] I have chosen instead to highlight the shared medialogical features of the crowd so as to attribute the proper weight to their recently alphabetized status.

Second, as the Italian state's experience with indoctrination during the First World War demonstrated, it is easier to inscript narratives of victory and defeat and sacrifice and heroism onto masses of men if they have only recently learned to read and write. In a recent reflection on digital simulations and analog subjectivity, N. Katherine Hayles draws a parallel between narrative inscripting and visual modes in evolutionary schemes, which is useful in understanding the importance of narrative for the recently literate at Fiume:

> Our sophisticated perceptual-cognitive visual processing evolved coadaptively with our movement through three-dimensional space, so it is no surprise that the creation of narrative is deeply tied with imagining scenes in which actions take place. . . . Articulated in this *lingua franca* of Western cultural perception, the images allow narrative to kick in with maximum force, for the action is "seen" in terms we can easily relate to our ongoing narrativizing of the world.[28]

D'Annunzio's creation of a national political style at Fiume with its flags, elaborate rituals, and dramatic symbolism was in no small measure indebted to his success in inscripting the past and future actions into a set of canonical stories that the crowds could suture back into a range of possible actions. The speeches and visual markers provided the favored narrativizations. To cite Mosse again, for the crowds to be "swayed and controlled," a dynamic had to be maintained— something that needed visual rather than merely verbal approaches."[29] It seems likely that the literate and semiliterate legionnaires would more easily attribute desires to D'Annunzio and their collective Fiume selves precisely because the speeches were crafted to ensure the desired kind of narrative interpolation. I will have much more to say about the narratives of Fiume momentarily, especially the mode by which sacrifice and salvation are married to a technologically inflected apocalyptic tone. What I want to mark at this juncture is the elemental power of D'Annunzio's speeches to inscribe narratives in a listening public in such a way that their future actions would correspond to expected behaviors, chiefly that "the degeneracy and death of Italy" would mean "its resurrection through heroic deeds."[30] Put differently, the speeches operate as recruitment procedures to the degree they inscribe their listeners with canonical narratives easily harnessed to the events unfolding at Fiume.

Alarmist Narratives: Saving the Patria

Not surprisingly, the most decisive narratives enlisted at Fiume are those in which a host of forces—principally Jewish bankers, Woodrow Wilson, and Francesco Nitti (then head of the Italian government)—threaten the *patria*. By dint of association, Fiume comes to stand for what Italy ought to be, had it not been for the "mutilated victory" proffered at the Paris peace talks: more virile, certainly more noble, and most importantly more attuned to an idealized notion of the fatherland. Thus in an early oration, D'Annunzio announces that the *patria* "is here. Here again one breathes in the heroic air, one again gasps in glory, one pulsates with delight, one is resplendent with keen desire."[31] The clear reference for the numerous *arditi* in the crowd would have been their experiences in the trench; indeed, the speeches go out of their way to associate the occupation with wartime ordeals.[32] The longer the occupation and siege last, the closer the *patria* is identified with Fiume. "The *patria* is something remote, solitary and occult, similar to the face of the Son of Man imprinted on the holy shroud. It remains a place of life and it is Fiume."[33] I will return to the Christian cosmology deployed in the speeches shortly, but of interest is how a heroic narrative is inscribed into a set of possible actions at Fiume. Initially, the *patria* will be saved as long as the *fiumani* hold out; in this way Fiume will reignite the fires "that were extinguished on the altars of the *Patria*," as D'Annunzio later writes in *Notturno*.[34] Inscribing the narrative of Fiume within the discourse of national heroic struggle anticipates and structures the crowds' reaction to the occupation, thereby presenting listeners with powerful possibilities for creating reflexive loops in which the environment at Fiume could come to resemble imaginatively an idealized form of *patria*.

Yet another narrative appears almost concurrently and grows in prominence as the occupation nears its ignominious conclusion: failing to hold the city will have no bearing on whether the fatherland will have been saved. This is because "the cause of the *patria* is given to us alone. We can save it by winning and we can save it by sacrificing ourselves. The perfect sacrifice is always a future victory." A heroic defense of the city gives ground to the performance of martyrdom in a language of sacrifice. The supreme sacrifice will mean remaining resolute even if that requires "burying ourselves in its [Fiume's] ruins, being burned alive in its flaming houses, laughing in the face of all danger and of the cruelest sort of death."[35] The name by which D'Annunzio marks the sacrifice demanded of the legionnaires is the "Città Olocausta, the city of the total sacrifice."[36] As Giorgio Agamben points out in the context of the Shoah, the term "holocaust" is a transcription of the Latin *holocaustum,* which is in turn a translation of the Greek term *holocaustos.* Its semantic history is Christian, "since the Church fathers used it to translate—in fact with neither rigor nor coherence—the complex sacrificial doctrine of the Bible." Used initially to refer to the sacrifices of the Jews in the Old Testament, "it is extended as a metaphor to include Christian martyrs, such that their torture is equated with sacrifice. . . . Christ's sacrifice on the cross is ultimately defined as a holocaust."[37] The introduction of the term thus ought to be joined in the speeches to attempts to reinscribe sacrifice at Fiume within a Christian narrative, in which the entire city is transformed into one enormous offering to an ideal *patria*; it functions as a potent "archaizing strategy for maintaining cohesion and consent" among the forces present at Fiume, drawing on the cultural history of many there.[38] That D'Annunzio will aestheticize their present and future sacrifice as a work of art, calling it repeatedly "the smelting of a statue," in much the same way that the First World War in a fit of Dionysian inspiration created its "one statue, one form of heroic humanity," is of course in keeping with much of his own writings from the same period, especially his war diaries and the memoir *Notturno.*[39] What is of particular interest and what most commentaries of Fiume fail to register is how the speeches construe sacrifice (and violence) aesthetically while concurrently reconsolidating Fiume as the ideal *patria* via a series of operations that might best be described as technologically indebted. Inscribing listeners with recognizable and canonical narratives borrowed from a Christological series is really only half the "Fiume experience." To uncover the other half, we need to turn to the question of an apocalyptic tone, wireless marconigrams, and deterritorialized perception. We will find the speeches both inscribing narratives and expounding a logic of prophetic unfolding that moves through a wirelessly produced delirium.

The Sounds of Apocalypse

Derrida's Apocalyptic Tone

The term "apocalyptic tone" as opposed to apocalypse or revelation requires an introduction. The expression originates in Derrida's reading of Kant's "On a Newly Arisen Superior Tone in Philosophy" and, in particular, in the procedure he adopts there for distilling tone from voice.[40] Most readings of the essay highlight how often the text foregrounds its own status as apocalyptic discourse and its striving to articulate, in Christopher Norris's words, "a language 'beyond' the analytic grasp of traditional philosophy, yet one that would (in some sense) keep faith with the need to resist all forms of obscurantist or irrationalist thinking."[41] My treatment derives instead from the association Derrida makes between the voice of the oracle and its preferred medium of transmission in the poetic. In order to elaborate the relation, it will be necessary to explicate those sections of the essay that place in relief the sounds of apocalyptic disclosing that are unmistakably present in the Fiume orations.

Derrida begins by identifying apocalypse with a scene of disclosure and the inspiration that follows at the sight of revelation, whether it be the secret part, the genitals, or whatever conceals itself. He is of course strictly following the etymology of apocalypse in "to unveil, disclose," which returns in his metaphorical understanding of gesture affording sight: "The question not only opens to vision or contemplation, not only affords seeing but also affords hearing/understanding."[42] Recapitulating the relation among disclosure, apocalypse, and writing, he addresses the effects of a reading that mixes the voices of the other in us (or of us in the other). These voices, the voice of reason and the voice of the oracle, differ in the relationship they enjoy with the ear. The former is sublime: "It orders, mandates, demands, commands without giving anything in exchange; it thunders in me to the point of making me tremble; it thus provokes the greatest questions and the greatest astonishment *[Erstaunen]*." He notes Kant's identification of the moral law with the voice of reason, and in a move in accord with his earlier work, he demonstrates how the voice of reason is attuned and tuned to the voice that hears and understands itself, but neither touches nor sees itself, "thus seeming to hide itself from every external intuition."[43]

For its part, the voice of the oracle takes up a position not opposed in some binary to the voice of reason but as its parasite. Where the voice of reason is untouchable, seemingly present to itself with its calls for duty, the voice of the oracle is all mediality. It demands that the ear be bent to the other's desire or pleasure—the oracle as other sends word of what it wants, addressing coded messages. A decisive passage follows: "The tone leaps and is raised higher when the voice of the oracle takes you aside, speaks to you in a private code, and whispers secrets to you in uncovering your ear for you, jumbling, covering, or parasitizing the voice of reason that speaks equally in each and maintains the same language for all."[44] Isolating Derrida's argument further, the voice of the oracle uncovers ears precisely because of its status as medium: were there no initial distance between ear and oracle—if the voice of the oracle did not give the distance proper to the ear—then no whisperings could be articulated. The voice of reason has no need of ears, that is where Derrida's argument pushes.

Matters are complicated, however, by the correlation Derrida draws between tone and oracle. If the voice of reason needs no ears to be heard, then it follows that the voice of reason has no tone to distinguish its various transmissions. This explains how reason "speaks equally in each and maintains the same language for all"—it lacks tone. At this point, Derrida undertakes to hear the tone of the oracle's voice. Beginning by locating its etymology in "the tight ligament, the cord, the rope when it is woven or braided . . . the privileged figure of everything that is subject to stricture," he links the pitch of tone to tension: "It has a bond to the bond, to the bond's more or less tight tension."[45] The etymology brings to mind a child's rudimentary telephone of two plastic cups joined by a string that doubles the bond producing the voice (the vocal chords). Depending on the width and the tension across the cord, varying pitches will be distinguished: the greater the tension, the higher the pitch.

Returning to the moment when the oracle bends the ear to its whispered sweet nothings, the tone "leaps and is raised higher" when the voice of the oracle begins to speak in code. Although Derrida nowhere suggests this, one assumes this is the case given the greater tension on the line between the ear and the mouth of the oracle. The oracle strings the voice of reason out or perhaps is already strung out by the oracle in an operation that untunes the voice of reason, what Derrida denotes in the German *Verstimmung* and what is translated in English as an "overlordly tone." "Generalized *Verstimmung* is the possibility for the other tone, or the tone of the other, to come at no matter what moment to interrupt a familiar tonality."[46] Accordingly, revelation does not primarily afford sight (though it is also that) but breaks into "a familiar tonality." It is, to use Avital Ronell's helpful formulation, the emergency broadcast system as tone of the other.[47]

To break into programming with an overlordly tone requires the prior approval of the authorities, who ultimately consent since they consider it to be a matter of life and death. Wherever there is a move from concepts to the unthinkable or the irrepresentable, Derrida tells us, a leap toward "the imminence of vision" and the unthinkable occurs, obscurely anticipating "the mysterious secret come from the beyond."[48] Those that have a presentiment of what is to come anticipate the moment of disclosure *(apokalupsis)* and thereby merit the principal accusation of mystagogue launched by both Kant and Derrida. By dint of an ability to see the future, the mystagogue, *il duce* as Derrida calls him (a descriptor to which I will return), initiates others into the mystery of the secret; his agogic function places him "above the crowd he manipulates through the intermediary of a small number of initiates gathered into a sect with a 'crypted language.'"[49]

Anticipating where Derrida's argument leads, it seems that those lifting the veil to disclose secrets are in fact mouthpieces of the oracle; whoever avails himself or herself of an apocalyptic tone interrupts a familiar tonality and makes the voice of reason delirious. *Il duce* takes on the apocalyptic tone in an attempt to reveal the truth, with tone functioning as the medium of some unveiling already underway: "unveiling or truth, apophantics of the imminence of the end, of whatever returns at the limit, at the end of the world."[50] Tone reveals truth, the imminent "just around the corner" that signifies that the end is beginning. Inasmuch as tone signifies the end, its function is "to seduce in order to lead to itself, or to the place where the first vibration of the tone is heard, which is called, as will be one's want, subject, person, sex, desire."[51] The connection between the mystagogues and an overlordly apocalyptic tone resides then not in some ontic notion of tone ready to use on the unsuspecting. Tone is not content; it is not what the mystagogue says that seduces but instead the vibrations the event of disclosing emits.

Two features of apocalyptic tone are noteworthy when discussing the events at Fiume. First, scenes of unveiling the truth may be magnified so as to garner attention among the listeners: martyrdom, for instance, could work well in infusing an overwhelming desire to understand that "bypasses human reason by being grounded in the direct disclosure through vision or audition."[52] The experience of the war with its millions dead demanded that an ultimate purpose be revealed, which would grant readers and listeners access to an overarching oracular perspective in which the real significance of the destruction could be unmasked. Hence the dramatic reanimation of war dead at Fiume, only to be killed again, fulfilling D'Annunzio's own "sinister prophecy" of sacrifice.[53] Second, an apocalyptic tone moves those who hear it since it is localized in a place "where the first vibration of the tone is heard." Although Derrida does not offer more details, he suggests that the apocalyptic tone is localized in a space constructed by the first vibration, which he immediately associates with subjectivity and desire. The introduction of an apocalyptic tone therefore recalls the succession of moments of hearing and their associated pedagogical spaces: a mother's mouth that instructs children on proper pronunciation, as Kittler in the German context has pointed out, or in a more modern setting the first vibrations of a tone heard over a wireless headset. I want to advance the position in what follows that *il duce* can, under the proper conditions, simulate primary acoustic moments and their concomitant space when the first vibration actualizes the creation of subjects. He can make new acoustic worlds rise up that free hearing momentarily and then reinscribe it within a different tonal register, providing newly constructed listeners with the means to recode its meaning. Because *il duce* pronounces and interprets the apocalyptic tone, he sets himself at the origin of discourse, which might account for what appears in Derrida's essay as a severe coupling between *il duce* and his minions reached via a crypto-poetics. The result is an apocalyptic tone that reconfigures the prior acoustic organization.

The Fiume Apocalyptic Tone: Addressing the Crowd

Given the Christological rhetoric deployed in the speeches and the narrative of holocaust that the speeches inscribe as canonical, it would be odd if the apocalyptic tone described by Derrida was not also generated at Fiume. Treating his exegesis of unveiling as paradigmatic, one wants to ask how the speeches set about creating conditions in which ears may be opened to acoustic events through an apocalyptic tone that in turn concretizes a certain kind of listener/agent at Fiume. The operations specific to generating the sound of disclosure at Fiume may be distinguished as follows, though they are not necessarily concurrent: staging Fiume principally as a series of acoustic events that require decoding for its mysterious contents to be revealed, a procedure akin to the wireless practices outlined above; recomposing these moments as a predictive history in which Fiume precedes and anticipates the First World War; and addressing the legionnaires as one entity in order to exploit the alarm their collective ear has registered in the apocalyptic tone.

For reasons that will be clear shortly, I want to begin with the question of address at Fiume before turning to the staging of Fiume as a wireless message transmitted to the crowds present. In his speeches and writings from the period, including *Notturno,* the orations' address clusters around the crowd as D'Annunzio's personal instrument of war that speaks with one voice and hears with one ear. In the July 24, 1920, oration entitled "To the Arditi of Fiume and Italy," D'Annunzio celebrates the "density" of the *arditi* and their unanimity of action, which the speech evokes as the basis for their singular collective address. The speech opens anecdotally: during the day's tactical exercises, someone was heard to observe that the *arditi* were assaulting their targets too compactly. This leads the *comandante* to reflect on their new style of warfare: "The cry was so unanimous that it seemed to arise from one giant chest, dominating the angry series of machine guns and the choral thunder of the howitzers. The hilltop was taken in a second."[54] Extending the metaphor to include the animal-like, the prowess of the *arditi* soon becomes a "grayish wild animal put together as if the bones, the muscles the tendons the nerves were assembled in the violence and the velocity of fulvous beasts."[55] The collective speed of the assault excites D'Annunzio but, more important, creates a unified address to which he can direct his appeals; hence the speech's destination to the "arditi of Fiume." Continuing, he announces:

> The unanimous cry that reverberated the other morning on the savage mountain is the sound of your true power. Unanimity is the breath of your action; it is the impulse of your dash against the adversary, against obstacles and toward meeting the future.[56]

The unanimity that results from bodies moving at a certain velocity becomes in the passage the mode by which the future will be engaged collectively. Tapping their reservoirs of loyalty to the nation by evoking the speed that constructs them as a single receiver, the speech then links D'Annunzio as commandant to their capacity to assault the future. "Now I am your *capo*, Flames of Fiume, Flames of Italy. And you must not listen if not to my voice." Naturally the speech exploits what it proposes as the genealogy for *arditi* construction: by addressing itself to them and placing D'Annunzio's voice as the medium for future commands, the speech gestures to a particular kind of message transmitted to and destined for a collective ear.

The function of expectation and with it silence across the orations follows directly from the adopted mode of address so that the speeches continually figure Fiume as a series of acoustic events requiring the prophetic negotiation only the *comandante* can provide. Thus, despite the torrent of words spoken at Fiume, the actualization of the crowd as "creatures of my spirit" and "a form of superhuman will" is the product of vigilant ear attuned to the message of Fiume.[57] And, of course, the repeated appeals to be vigilant reinforce the message—to wait for the word that the sacrificial moment has arrived—as does the obsessive recalling of previous attentiveness during the war when "the attention and the waiting were in every drop of our blood just as the divine presence of the *Patria* was in every drop in the sea."[58] The speeches go out of their way to link vigilance to acoustics: in one of the final speeches, "Taking Leave from the Tombs," D'Annunzio sketches the previous day's pilgrimage to a nearby military cemetery and the silence of a city that "waits to hear, just as the anxious women placed their ear against the ground in order to hear the rumble of the March of Ronchi [the initial march that led to the occupation]."[59] All were quiet, he notes, and "one saw that for them silence was an element of placing in relief and expression."

Jeffrey Schnapp has insisted on the quality of the silence imposed at Fiume as an "immobile gestuality, monumental and commemorative," which he affirms as the virile alternative to writing, though not without paradox as it hovers on the threshold of action.[60] While certainly true, we are in a position to provide more details on the role of silence vis-à-vis action. The Fiume legionnaires are continually figured on the verge of movement, yet remain immobile and silent, because only in such a way will they be ready to perceive the apocalyptic tone announcing that the *patria* is in danger. Perhaps one should speak of a silence productive of a tone that announces that the end or the beginning is about to commence. Indeed, the less the *comandante* speaks, the greater the sense of anticipation created, which the same speech confirms: "Yesterday I could not speak, I did not know how to speak. I carried silence as one carries the revelation."[61] The speeches urge forward silence tautologically as the medium by which revelation is revealed and thus introduce a way of listening not previously available to the multitudes gathered at Fiume.[62] By positioning D'Annunzio paradoxically as the silent agent of disclosure, the speeches frame revelation as originating and reaching completion in the figure of the commandant.

The Fiume Apocalyptic Tone: Disclosing the Patria

A side glance at *Notturno,* D'Annunzio's account of his convalescence from a wartime biplane accident, shows that silence produced by adopting an apocalyptic tone is pregnant with the sound of disclosing the *patria*.[63] In those pages devoted to the vigil D'Annunzio keeps over the mutilated corpse of Giuseppe Maggiora, the text suggests that the chief characteristic of the heroic body is an ear that keeps vigil: "The nude hero rises again from the human crater. It sculpts itself anew in the mud of its origin. . . . There is a convulsed and bloody body; and in all that palpitating mutilation, there is a vigilant ear, there is an apex of a soul that hears and longs to hear."[64] What principally distinguishes the hero from the mud is the isolated ear attached to a traumatized body: a message addresses itself to the ear that hears and desires to hear. The hero gleans that the message originates from the *patria* in moments of darkness, when signs illuminate the night, inscribing the body in fire and pain. "I hear the name of the *Patria*; and a great tremble moves through me. I hear again the name of the *Patria*; and the same trembling passes through my marrow."[65] The passage posits an economic inscription of the body as opposed to a reading of the signs as the mode by which the *patria* reaches its target.

As the memoir makes clear, inscribing crowds occurs in similar fashion, when the loss of an eye is made analogous to the procedures by which the crowd comes to hear the *patria*. In the first instance, D'Annunzio's near-blindness (one eye lost, the other eye covered with gauze) is made the means by which acoustic events penetrate the body. "I have never suffered the human voice as I have this hour. The querulous confusion is like an instrument of torture that operates at a distance. My living body is contorted and drilled in various ways, with focus and with violence, according to the accents I hear." Transformed into an unwilling witness to cinematic visions thrown up by his unconscious, he no longer moves but registers, hearing the "ticking of the clock, that now seems to me to be a termite in my ear, now distant like the quivering of a star." The teletorture he describes provides him with a new notion of temporality linked to heightened acoustic registration. "New notion of time. Struggle between the live image, continually created by memory [*ricordo*] and the immobile body." On the one hand then, acoustic perception is deterritorialized when eyesight is lost, with sounds

A half-blind D'Annunzio convalescing. Reprinted by permission.

assuming the power to instigate intense visions that "force me now to feel the same pity and the same anguish that assaulted me earlier." These visions in turn are reterritorialized as apocalyptic revelations of a past that doubles for the present: "The past is present in all its aspects, in all its events." Moments become eternal as the bed spins underneath him, leading to the fall that "has no end." In *Notturno* immobility and near-blindness guarantee D'Annunzio access to revelations of the future. The result is paradoxically both a heightened sense of perception and the creation of a mouth that no longer articulates the human word. Thus: "I feel that my hands have become more sensitive . . . something unusual, which resembles a *chiarore affluito*," while "the mouth opens and it cannot speak the human word."[66]

Notturno also invokes lack of sight as the instrument that bends the crowd to the will of the commandant. When recalling, for instance, the speech of May 17, 1915, in Rome, just as Italy was preparing to enter the war, D'Annunzio writes:

> The theater is like a deep smelting furnace. I see the metal cool. The boiling statue of the crowd becomes solid, it stabilizes itself in a threatening relief, it is like a mass that crushes. It [the crowd] no longer cries out. It listens. It doesn't breathe. It listens. Every syllable penetrates the perforated skull and remains there unmovable. It is an atrocious work of the hammer operating on a silence that resists.[67]

Note here the emphasis placed not on the words spoken but on the "syllable" that embeds itself. In order for the process of crowd construction to occur, however, the crowd must first be recognized as a crowd, its address as crowd stabilized, what the passage foregrounds aesthetically as the move from liquid to metal. Then it must be immobilized—this after it no longer "cries out" or breathes. Blocking speech isolates the ear, creating the necessary silence for the hammer's work to begin. Curiously, nowhere does the passage gesture to the visual in the limited exchange among oration, speaker, and crowd. Moreover, D'Annunzio nowhere offers details on the process by which the crowd shakes off its verbosity to become the taciturn acoustic registrating machine that offers little resistance to the rhythmic pounding of syllable on syllable. At this juncture, what emerges in our reading is the staging of the crowd-commandant nexus as principally an acoustic one; the impression being that the form of listening the crowd practices is contingent on not speaking.

Indeed, the Fiume orations repeatedly mark listening at Fiume as both the principal means by which the crowd is distinguished from other less virile formations and the necessary condition for heroic action. In such a situation the role of the *comandante* is to reveal to the crowd what it desires, not by affording them sight, but instead by speaking to them in code that breaks up what

Derrida might call a "familiar tonality." So in the early moments of the occupation, "everyone vibrated, all were ready. It seemed that everyone already had in their souls the same dream, and I was nothing if not the improvised guesser and interpreter." And later: "I was not anything except the interpreter of their closed cantos. If the true poet is he who does not walk if not in his own blood, I here with circumspection speak to you in my language of the poet to liberate your canto that is in you and the courage that rings in you."[68] The speaker/ poet anticipates the crowd's reaction to the disclosure of their own future as his agents, and so takes on an apocalyptic tone to the extent that an unveiling of truth is now underway. *Notturno* frames crowd construction at Fiume as an event of language or, better, an occasion of inspired speech in which a possible future of heroic action is imagined and performed in the memoir and by association in the speeches themselves. The future is reformulated and reassembled to correspond with one more in tune with sacrifice.

So too the past, which in the speeches and *Notturno* is reassembled as a reservoir of Italianness that intrudes on the present and future. Thus D'Annunzio says that "we want to feel our Italianness *[italianità]* throughout all time, in all the space of our ancestors, even to the dusk in which our seas began to be illuminated by the beauty of our coasts . . . even to the fecundation and the conquest of the most remote future."[69] The implicit justification of Italian imperialism is obvious here, but more important for my purposes is how the passage organizes the past as an instance of *italianità* that respects no temporal or spatial constraints. Fiume commands in a message launched out of the past: "From the depths *[fondo]* of the centuries it commands the future, like the gesture of that leader *[condottiero]* that has returned."[70] Fiume is not simply the symbolic equivalent of future sacrifice nor the present moment of occupation but a command that reaches from the past into the future. Generally one associates the collapse of temporal and spatial confines with the effects of technology—McLuhan will quickly come to mind as will any number of Italian commentators—so the passage suggests a relation between Fiume and a presumed deconsolidation of space and time.[71] Although it is not yet clear how the speeches authorize Fiume as the means for escaping from these constraints (aside from recognizing an essential *italianità*), what does seem clear is the trajectory from certain past to undeniable future moves through the shrinking of space and time.

To summarize: the Fiume orations along with *Notturno* apply apocalyptic rhetoric and imagery when reassembling the past and future. They do so by stabilizing a collective address in the crowd, exploiting the unanimity that results from bodies moving at accelerated velocities. Concurrently, they posit silence as the medium by which the crowds are actualized as instruments of the commandant; vigilance the purview of an apocalyptic tone that creates alarm in its listeners for the fate of the *patria*. What remains to be explored is how the speeches frame and authorize the commandant's transformation into a prophet with access to the future.

Marconi, the Wireless, D'Annunzio

Framing Fiume

The moment described earlier when an apocalyptic tone reconsolidates the confines of time and space is one I want to examine more closely in the pages that remain by commenting in detail on the speech, "Greeting to Guglielmo Marconi upon His Arrival to Italian Fiume." One of the shorter discourses spoken at Fiume, taking up no more than six pages, the speech is at first glance typical of the many that precede it, especially in its apocalyptic imagery, in its economy of words in tune with the D'Annunzian project of embedding sounds, and in its genealogy of the past as inevitably generated by and moving inexorably toward Fiume and its holocaustal fires. In its evocation of the 1915 meeting between Marconi and D'Annunzio, the speech calls to mind D'Annunzio's war diaries. Indeed, on closer examination, it would seem that a great deal of the speech originates in an until recently unpublished diary entry of D'Annunzio's from August 1, 1915.[72] What sets the speech apart, however, from the ninety or so speeches D'Annunzio gave over eighteen months, is its stark marking of the procedures for constructing acoustic subjects that Marconi's device and its associated practices of hearing and writing will authorize. Equally significant is the authorization wireless technology provides the commandant as he attempts to hear in Marconi's body and wireless apparatus a future of sacrifice. In different words, resorting to the genre of wireless writing will provide D'Annunzio with access to the significance of Fiume and the First World War. For these reasons the speech is one of the most important (and overlooked) documents of modern Italian culture and so deserves close scrutiny.

How then does the speech appropriate wireless technology to frame the events at Fiume? The short answer is through an acoustic identification between Fiume, Marconi's body, and a mysterious message that requires decoding. After greeting and thanking Marconi for his seemingly unexpected arrival, D'Annunzio draws a complex relation among Marconi, vibrating sounds, and a mysterious message addressed to those present at Fiume. "Today docked at Fiume, disembarked on the furthermost shore of heroic beauty, does he [Marconi] not seem to carry in himself all the vibrations of the most mysterious message, oh citizens of the City of Life?'"[73] The passage stages the arrival as an acoustic event, marking Marconi's body as the source of all vibrations emanating from a mysterious message that Marconi himself strangely is carrying. Although not stated directly, the suggestion is that the message is addressed to the "Citizens of the City of Life" at Fiume given Marconi's appearance with D'Annunzio on the balcony. The contents of the mysterious message remain undisclosed until the speech's final moments and, even then, in typically apocalyptic fashion, are never fully divulged. By postulating an identification between message and Marconi, one notes that the speech positions D'Annunzio as its decoder: he will tell his listeners how to listen to the message that comes from beyond (with the arrival of Marconi and his ship, the *Elettra*), and so will reveal not merely the message but the relay and origin of the message itself in the figure and body of Marconi. The oration's tone is initialized in decidedly apocalyptic fashion, for as it promises to make the Marconi vibrations heard, it simultaneously offers itself and D'Annunzio as the agents of disclosure. Thus, the text has the contents of the mysterious message and the emanating vibrations take up positions as markers of an apocalyptic tone inasmuch as they indicate an acoustic unveiling to come.

By joining a monumental Marconi body that generates vibrations to the Fiume occupation, the speech deserves a special place in the halls of Marconigrafia. The reader will recall how easily the Marconi body took on the characteristics of wireless equipment in the Dunlap and Landini biographies of the first chapter; D'Annunzio is little different to the extent Marconi is transformed into a vibrating medium for the mysterious message. Moreover, the passage registers a great deal of mystification surrounding the wireless invention—what precisely one wants to know is the relation between Marconi and his transmitting

Severi's rendering of Elettra's *message. The caption reads, "The spark of Italian genius." Courtesy of the Archives of the Fondazione Guglielmo Marconi, Italy.*

equipment—especially in the move that identifies the "most mysterious mes-sage" with Marconi. As Avital Ronell has tirelessly pointed out, "all forms of identification that are structuring emerge from a trauma, or from a reserve of what is missing," so we will want to determine what shared reserve of lack joins the message with Marconi.[74] In the case of the inventor, we know the enor-mity of his lack, spelled out in the last chapter in wired mouths, braces, and unauthorized patents. The same holds true for the "most mysterious message"; some structural lack (of transparency, of closure) neatly accounting for its sta-tus as mystery. I will return to this question shortly as the speech concludes.

Marconi's Mouth

But it is not simply a matter of D'Annunzio hearing and decoding the "most mysterious message," since the speech reconfigures Marconi's lack as a monu-mental mouth forever on the verge of disclosing the contents of the message. Here the text's transformations of the function of Marconi and his body are instructive for understanding the presumed analogy between Marconi and Fiume. First, Marconi is "the genius of Italy diffused universally with the swift-ness of stellar light" who comes to Fiume to "amplify indefinitely the sound waves of the voice of Fiume" and to "arm with speed our defiance, our re-sponses, our protests, and all the affirmations of our rights, of our courage, of our tenacity, all the appeals of our pain and our ardor."[75] Marconi supplements the voice of Fiume by his mere presence at Fiume, amplifying and quickening the words that together constitute its protest. The speech traces the outlines of a rudimentary form of wireless transmission in the figure of Marconi, attaching a mouth to the site from which a vibration emanates, making it circulate over ever-increasing distances. Essentially, the speech stages the transmission as a scene of ventriloquism since the site of the vibration is Marconi's body and not primarily a voice (Marconi's, D'Annunzio's, or Fiume's).

A Marconi capable of traversing the stars shoots back to earth on the fol-lowing page when D'Annunzio again evokes the mouth of the inventor during the Great War: "That thin and sensitive mouth that sometimes has a smile of sweetness and infantile ingenuity, was closed by the hermetic seal of the secret." D'Annunzio pleads with him to break the seal and speak his piece. "Oh Wizard, is it true that you have succeeded in seeing across the seas, with a sight sharper than that attributed by the ancients to the lynx?" Marconi then "gestured like Arpocrate, the son of the mysterious Iside, putting his finger in his mouth."[76] From universal omniscience to petrification in a monument that continually invites one to keep one's mouth shut (while reflexively shutting its own), the speech figures Marconi's body first as being without one, reconstituting it soon after as an enlarged mouth sealed in place by a finger. In this regard one recalls the wired mouths of the first chapter, painfully worked over as one-way transmission sites.[77]

Wireless Practices

With Marconi's capacity as medium guaranteed rhetorically in the speeches, we will want to raise a number of questions. How does the speech cite, reproduce, and verify the forms of transmission and reception in operation at Fiume as wireless practices? What precise meaning should be given to the formulation Marconi speaks as the voice of Fiume while being simultaneously sounded in it? Last, how does such a formulation function in terms of generating an apocalyptic tone? In answer to the first question, note that both the wireless and Fiume occasion access to a past suffused with the possibility of future action. It is one May during the First World War and D'Annunzio, accompanied by Marconi, is touring the wireless station of Centocelle near Rome. "It was one of those Roman hours in which one feels that there where there is death, there is also resurrection. It seems that the swirling winds, raising up the ashes of the sepulchers, changed them into seeds of the future."[78] D'Annunzio's recollection mimics one of the principal effects of the wireless transmission—its apparent ability to sow the seeds of the future when it broadcasts, reaching beyond the grave and the past to connect life and death. The passage echoes one of the principal characteristics of the Fiuman experience itself, namely, the reassembly of the past in terms of an illuminating sacrificial future. Given their respective capacity to reach into the past, it is not surprising then that both Fiume and the wireless also broadcast over great distances. Indeed in a speech given a month before Marconi's arrival, the message of Fiume is transformed into a message as vast as the earth itself, "running from Dalmatia to Persia, from Montenegro to Egypt, from Catalonia to India, from Ireland to China, from Mesopotamia to California. It embraces all oppressed races, all contrasting beliefs, all suffocated aspirations, all failed sacrifices."[79] The wireless, with its capacity to transmit over ocean and all sorts of obstacles, as witnessed in the previous chapter, is a natural analogue for the kind of message D'Annunzio hoped to send from Fiume to the farthest reaches of the globe.

The speech continues to elaborate similarities between Fiume, Marconi, and the wireless in its ensuing description of the details of transmission. Traveling together in a "fast coach, each by the other's side," D'Annunzio commences his soliloquy of the wireless while Marconi remains silent with his polished saber between his knees.

> And I called forth with my imagination the immense radiotelegraphic network that he had spread around the world, the measureless game of invisible

waves, which in that hour were spreading out to all the blood-stained lands, carrying the calls and the responses, the announcements and the entreaties of men, the cry of danger, the message of victory, the admission of defeat.[80]

If Marconi's body is not simply the site of vibrations but also the container of everything possible ("All the possibilities were in him, as in that other magician, as in Leonardo da Vinci," D'Annunzio will say soon after), so is Marconi's radio-telegraphic net allied with possibility.[81] It counts among its abilities a singular capacity for carrying a variety of messages—appeals and responses, cries and announcements, and messages of victory and defeat are all transmittable over the Marconi wireless. One notes as well that the evocation of the wireless does not assume a clear-cut destination for the wirelessed messages but rather suggests an enormous circuit in which messages, cataloged in terms of question and response, circulate continuously. What unites them is their mode of travel through the invisible waves that carry them above a bloody earth. In similar fashion, the message of alliance among the excluded and forgotten launched at Fiume circulates far and wide, carried along by its universal power to touch souls. Indeed, even from before the occupation, D'Annunzio's orations posit a message that echoes across the earth. In the speech "The Wings of Italy Are Freed" (July 9, 1919), those listening to D'Annunzio "hear the appeal of the Arabs and of the Indians oppressed by those 'in the right' who hold our Malta and have torn Fiume from us."[82] By December 1, 1920, in the closing weeks of the occupation, the cry has become a message that "passes through the fog of the rainy evening, passes above the anxious Vigil, and reaches finally there in the dead breast of Carnaro, where the few battered but tenacious survivors of *italianità* today seem to transform the wheeze of their sublime agony into a roar."[83] Marconi's arrival—and by association the wireless message—frame the speech's concern with transmitting a message of liberation and sacrifice to as many as possible as well as verifying D'Annunzio's prophetic claims to knowledge.

The Case of the Austrian Marconigram

Next the speech reproduces wireless acoustic practices as the technological equivalent of new forms of hearing at Fiume. As their tour of the wireless stations continues, D'Annunzio and Marconi inspect more closely some of the inventor's recently installed equipment. The result will be to generate an apocalyptic tone in both the wireless transmission and in the Fiume message. Marconi, figured once again as "a taciturn soldier closed in his iron discipline," hooks himself into the wireless station:

> He examined the equipment with a familiar gaze, he touched it almost caressingly with his hand, as enchanters handle their fascinating beasts. The

immense cosmic energy spoke with that calm man; it possessed a language he understood just like the speech of his child.[84]

The passage describes the move from a primarily visual acquisition of the wireless ("a familiar gaze") to the tactile ("he touched it almost caressingly") to the auditory ("the immense cosmic energy spoke with that calm man"). Although it is admittedly difficult to hold together the image of a cosmic energy that speaks to Marconi in the tongue of a child, the speech evokes a realm of primitive wireless cryptography in operation at Fiume. It identifies the sender of the message addressed to Marconi as the cosmos that speaks to him through his device (or what remains of the apparatus as D'Annunzio continually elides its materiality in favor of its ideal auditory features in capturing the voice of cosmic energy). An inventory of wireless condensers and Leydian batteries follows and the passage ends with Marconi's arteries coupling him to the wireless; they appear to "obey the secret will of only this man; they appeared to respond to the beating of his arteries."[85]

With Marconi's preliminary examination complete, both bend their ears close to the receiver, each man balancing himself with his saber "so as not to let it drag on the pavement." Listening attentively, they hear signals transmitted "by the most remote stations, revealed by the quality of their tone; a message from France; a message from England; a message from Russia; a message from America."[86] Suddenly, a subaltern of Marconi says in a sufficiently low voice to interrupt the tonality of the various transmissions: "An Austrian marconigram!" What D'Annunzio and Marconi pick up and how D'Annunzio decodes it warrant substantial citation.

> And he took off the earset placing it on me. My great friend and I looked at each other with a start, united by the same thought, by the same emotion, with the same thrill in the veins and marrow, with the soul leaping far away, in a concordant rush toward a frontier, lands in which our brothers were fighting, toward lands already wet with the purest Italian blood, already transfigured by the highest Italian passion. Oh bare Alps of Trento! Oh sea of Trieste, the martyr! Our imperial Aquileia, too long profaned by the barbarians! Our Dalmazia, narrow as the edge of a toga, but a Roman toga! Our Fiume, our Fiume, oh how desperately our Fiume!
>
> Yes, comrades, yes brothers, this is the cry that crosses all space.
>
> Don't all the aerial antennae scattered over this vast world vibrate with your acclamation?[87]

I want to localize three moments of particular interest here for the reproduction of wireless practices and then comment generally on its importance for reconsolidating Fiume as a relay for future and past transmission. First, note

the spatial and temporal characteristics of the transmission. The wireless trans-
ports D'Annunzio to the frontier where the most important battles of the Great
War are being fought simultaneously. The speech—similar to the oration described above in which *italianità* guaranteed both time travel and travel out
of time—represents the frontier as an auditory space in which D'Annunzio
and his listeners (*marconisti* and the Fiume legionnaires fused together in a
progressive present) hear a directory of battles fought during the war: Trento,
Trieste, and Dalmatia are transmitting on the wireless under the cover of Fiume
to D'Annunzio, who picks up the signals in the past, heard at a Marconi station near Rome, which is present at Fiume. To complicate matters, however, the
signals are arriving from the future, since the battles have yet to occur, to a receiver located in the past (one assumes 1915 though the date remains unclear),
having been sent in the form of "an Austrian marconigram." The final lines
reveal the signal's contents as containing the same message, "Our Fiume, our
Fiume, oh desperately our Fiume," the cry that overcomes spatial and temporal
restrictions. The reproduction of wireless effects in the Fiume message is striking. The shrinking of temporal and spatial distances that mystified so many
at the turn of the century is noted standard operating procedure for the wireless. Moreover, by attributing an immediacy to the wireless and to Marconi, the
speech recalls some of the magic shows Marconi performed at Stonehenge and
elsewhere (see chapter 1). Here D'Annunzio puts on the earset and has an auditory hallucination that brings the past to life. The magical qualities the speech
invokes when detailing wireless practices are not incidental since the grafting
of magic onto hearing at Fiume not only authorizes Fiume's transformation
into a truer and more virile Italy via the wireless but will also assist in establishing D'Annunzio's prophetic credentials.

Second, Marconi turns D'Annunzio on and onto the wireless, fascinating
his ear in the process. D'Annunzio begins to sense "a thrill in the veins and in
the marrow," which the speech inflects as the inevitable effects of a technology
capable of reaching into bodies. The process continues in the series of transmissions that follow, in which the wireless gradually increases the dosage: first
he hears the French marconigram and then the English, which enables his ear
to distinguish the tone of the signals, signals whose very site of transmission,
Fiume, is its message. The significance of the last formulation will be found if we
judge it to be essentially a leap toward what is only represented with difficulty
or what cannot be sufficiently held together in the mind. To the degree that the
sender and receiver of the message are undecidable, the passage registers the delirious effects of the wireless transmission. Some unknown future already past
transmits on the Marconi wireless, from a future Fiume to a past D'Annunzio;
the wireless revealing to D'Annunzio the message that cries out Fiume since
he hears the message through the earset. By carrying the signal of Fiume, these

marconigrams predict the future by calling out the name of a future transmission and a future marconigram. Perhaps it is not so much D'Annunzio working the levers of apocalypse in the passage but rather of him availing himself of the possibilities of wireless transmission in order to fix the moment of revelation. In order to hear the message before Fiume speaks itself, the legionnaires require a device or a relay attuned to the future capable of registering presentiments of the future. To the degree D'Annunzio takes on the task of deciphering Marconi and the wireless transmission, he positions himself as their chief decoder.

If tone is the tension that joins ears to the mouth of the oracle as Derrida argues, then the tone of the passage consists not of a content to be assigned to the name Fiume. The speech suggests as much on the next page when D'Annunzio identifies Marconi again with Fiume: "Well, if this Magician has his secrets, Fiume also has a secret."[88] There can be no full disclosure of Fiume just as there can be no full disclosure for Marconi. Yet the formulation is paradoxical because the specificity of the device is met by the dispersed addresses of the transmission—do not all the world's antennae vibrate with the marconigram sent from Fiume? The interlacing of destinations in the wireless transmission corresponds therefore to an unmistakable strain of delirium that finds expression in the Austrian marconigram vibrating on all the world's antennae. Essentially then the speech comments on the means by which the message of Fiume appropriates an apocalyptic tone that the wireless generates, which is really another way of saying that the wireless transmission speaks Fiume through Marconi's mouth. Inasmuch as the wireless continually postpones divulging the contents of the Fiume message, the tone is apocalyptic.

To a degree unmatched by the telephone and the telegraph before it, the wireless lays down lines of communication to the voice of the oracle by criss-crossing and interlacing destinations and addresses and hiding its operations under the cover of universality.[89] The wireless works both as a mouthpiece for the voice of the oracle and an earpiece to the other, placing its human listener in a certain position vis-à-vis the other's transmission. The wireless has already programmed what one will have heard, not in some hermeneutics of the transmission, but in the nonarticulated promise of the anticipated apocalyptic tone. In the same way do the speeches impress on their listeners the importance of remaining vigilant for the moment when the commandant will decode the message and sound the alarm. Both the wireless and the commandant design ears to distinguish the apocalyptic tone of their transmissions, which is manifested as an Austrian marconigram that signals Fiume. Forever promising to reveal the tone of the other, the wireless and the commandant maintain an apocalyptic tone.

In other words, the Fiuman experience, with its visual markers of the Italian sacred and exposition of decoding, operates similarly to a wireless transmission to the extent that both are structured paradoxically as revealing secrets or, better, a secret that opens ears and mouths to the mystery of transmission. Fiume promises revelation, but as it never discloses itself, it defers the moment when its truth may be spoken. Fiume becomes the site of an originary vibration that streaks like a lightning-fast Marconi from transmitter to antenna and back again, only to escape a final signifying moment. It functions as a place without location from which to get somewhere else, a relay station and a translation of the signal traversing the ether.

Marconisti at Fiume

In his discussion of apocalyptic means and ends, Derrida speaks of seduction and the promise of a strong signal whose origin is beyond doubt. Marconi's early, obsessive experiments may be read in this light: the weak "S" patched up into a strong "S" assures both sender and receiver of their positions in the transmission. Not surprisingly, D'Annunzio promises the same in the speeches because "the will of all free men in the deceived and vanquished nations must speak through the mouth of only one."[90] It is important to recall, however, that no marconigram, whether it be a coded Fiume or an "S," could have been received had not ears attuned to the tone of its transmission been capable of distinguishing it from others. This is true both of Marconi's subaltern, who relays the marconigram to Marconi, and D'Annunzio, who decodes the Austrian marconigram into "Fiume, our Fiume!" Put differently, both the wireless and the mysterious message of Fiume need agents of disclosure and acoustic specialists for their respective tones to be made out. The wireless gets a group of decoding experts who can distinguish the respective tones of a marconigram and whose hands can inscribe messages as marconigrams. Fiume has D'Annunzio as the principal agent of its strong signal. The next section focuses on how both program their operators, including D'Annunzio, for the ends of constructing a ritualized sect of initiates trained in the art of cryptography.

That the primitive wireless was a dictating machine that required a human interface for writing was plain in the first chapter when I described the first dictations over the wireless.[91] For D'Annunzio to hear the wireless message transmitting Fiume, the signal must first be decoded and then written down. The speech details the conditions necessary for the decoding in another of D'Annunzio's wireless ruminations with Marconi:

> For this reason we were Latins and Italians, full of the past, but thrust toward
> the future, capable of reliving all the memories and capable of renewing
> in ourselves all hopes, men of yesterday and men of tomorrow. We could
> bend down to read the inscription on a worn-away tombstone, there near the
> cemetery of *Equites Singulares*. We could catch in our brain the symphonic
> agreement between the transmitting and receiving station. We could guess,

using the sharpness of the ear, the origin of the transatlantic message. We
could ascertain the exact discourse of the receiver. We could evaluate the
sound produced by a series of oscillating groups created by a transmitting
disc six thousand kilometers away![92]

To hear the wireless message, D'Annunzio and Marconi must not only be attuned to the frequency on which the transmitter and receiver are working, they
must guess the provenance of the message "using the sharpness of the ear." The
passage confirms many of the features of the first *marconisti* in its emphasis on
guesswork and acute hearing as the necessary building blocks for constructing
a wireless operator. Despite the seeming naturalness of the process sketched,
there is little natural or common about picking up the signal: these *marconisti*
are in an intuitive relation to the mystery of this wireless message, trained to act
as intermediaries between the wireless and the transmission's destination.

The point I want to urge issues from the above instructions. *Marconisti* occupy the role of the initiated insofar as they have their own crypted language
and ritualized practices. They are not first and foremost agents of action but are
instead bodies trained to hear and act in the scene of wireless dictation and the
commands directed to them from the wireless authority.[93] In this regard, consider the placement of the *marconista*'s body between transmitter and receiver.
Solari relates that "at every interruption in the working of the Morse machine,
due to an irregular stopping in the relay, I would clasp the coherer on its sides,
having wetted my fingers with my tongue (as I learned from Marconi), so as
to close the circuit by using the resistance of the human body." The body itself
closes the circuit between transmitter and receiver in the primitive wireless,
literally forcing the signal through it, which explains the compulsory "Lick!
Lick!" Marconi was so given to when instructing his Marconi wireless operators.[94] The particulars of D'Annunzio's speech to this point confirm the ritualistic and cryptographic practices: a cosmic energy organ transmits to Marconi
near Rome in an infantile language only he understands; the anonymous wireless operator initiates D'Annunzio into the mystery of tonal marconigrams; and
finally D'Annunzio practices cryptography on the vibrating Fiume message. In
the wireless Fiume transmission, the mystagogue pilloried by Kant and Derrida
is not simply *il duce* bent on drawing crowds to himself but a mode of transmitting that leads users to wait indefinitely for the signal that will reveal that that
end has arrived.

The Wireless Instruments of War

Yet seduction through strong signals only partially accounts for the aims of
the wireless and the mysterious message of Fiume. Of singular importance is
the fact that both Marconi and D'Annunzio hear the message of Fiume at a

moment when the Italian army was facing one of its darkest moments after the Caporetto disaster. The context for this marconigram in particular is the forced alphabetization and indoctrination campaigns undertaken by the Italian state discussed earlier. Another principal aim of the marconigram will thus be the constitution of future subjects willing to die for Italy. As has been the case to this point, D'Annunzio's speech models our reading of the historical significance of the wireless for events at Fiume:

> We were two soldiers of Italy. His science and my poetry had become instruments of war, forces of combat, promises of victory. In the beginning there was Action; in the end there will be Action. This was the faith of his science, this was the faith of my poetry; it was this poetry that carried me to Fiume, to the Ronchi cemetery. Our every thought was born and developed with the rhythm of the will. If science has a universal power and if poetry is destined to touch all souls, we felt like men of our land, tied to our soil, tightly united to our race, devoted to one idea, soldiers of the one Italy.[95]

D'Annunzio reshapes his poetry and Marconi's wireless into instruments of war, vectoral forces drawing their charge from the principle of action: both poetry and technology command soldiers to act. In one sense, D'Annunzio is ventriloquizing Marconi's mother on the nature of the wireless invention when speaking of the power of poetry, which is to say that he recognizes that the wireless and his poetry demand forced marches from their listeners. But how does D'Annunzio's poetry—we will want to delimit the kinds of poetry in question considerably—demand the same forced marches from "the souls" it touches?[96] The short and difficult answer is by creating subjects willing to obey orders. D'Annunzio's poetry is a tool designed to meet the exigencies of war, which is not really surprising given the reading habits of the recently literate in the Italian trenches. That Italian soldiers were less literate than their English-poet counterparts does not alter the importance poetry, D'Annunzio's particularly, had for them. Indeed, soldiers have always required their medium-induced hallucinations to carry out commands, whether they be Italians in World War I or American GIs in Vietnam playing John Wayne war movies in their heads.[97] The move from traditional to technical warfare, with its hard-to-locate commanders bunkered in, meant that "the traditional rhetoric of presence and persuasion would be replaced by technologies of telecommunication and control."[98] D'Annunzio's call of poetry to arms is simply the recognition that both telecommunication and poetry serve not only to motivate troops but to construct subjects by opening their ears to the *patria*.

The identity D'Annunzio evokes between his poetry and Marconi's science, however, complicates the scenario considerably. Both have a universal power destined to touch every "soul." In other words, D'Annunzio's poetry and Marconi's science construct subjects able to pick up the message that will link them to their country and the dying that will be required of them. This D'Annunzian network of wireless connections will of course require the kind of invisible lines that mothers everywhere laid down for the state, the same mother-son relation that Heidegger so knowingly describes in *What Is Called Thinking*: "She will convey to him what obedience is. Or better, the other way around: she will bring him to obey. Her success will be more lasting the less she scolds him; it will be easier, the more directly she can get him to listen . . . not just condescend to listen, but listen in such a way that he can no longer stop wanting to do it."[99] Marconi's equipment and D'Annunzio's poetry both utilize the same lines set down by mothers everywhere and described so acutely by Heidegger. If both were to stand any chance of reaching the kind of public they deserved, that is, potential soldiers of Italy, they had need of the kind of invisible wires only mothers could put down. All done so that the wireless receivers of World War I, once called infantry, would be unable to put down D'Annunzio's poetry or miss the marconigram meant for them.

I noted above how often the Fiume speeches define universality in terms of the wireless's capacity to target entire countries with its signal; the idiom of the wireless and poetry rings with the sounds of commands being given over great distance and time. Indeed D'Annunzio's rhetoric continually stresses the immediacy with which his words, both at Fiume and elsewhere, will reach his listeners, the immediacy of the past present to Marconi and D'Annunzio in a passage already cited, "full of the past, thrust towards the future, capable of re-living all our memories and capable of renewing in ourselves all hopes, men of yesterday and men of tomorrow."[100] D'Annunzio's Marconi oration foregrounds what the other speeches implicitly suggest is the importance of Fiume, name-ly, moments when the past is simultaneously present with a wireless future: ancient Roman roads and the inscribed tombs of the Roman Republic exist alongside "the old iron towers, the antennae, the airplanes, the new sublime acropolis."[101] Sustained by the authorization Marconi's invention provides, D'Annunzio too overcomes the boundaries of time and space in that he hears the future Fiume signal in the past authorizing his future status as relayer at Fiume and prophetic voice of the oracle.[102]

Consequently, the wireless asserts an immediacy of transmission that fasci-nates D'Annunzio. As we shall soon discover, he was not alone: many Futurists and budding fascists were either spooked or excited by "the extension of per-ceptual powers that comes with radio hearing, aerial viewing, and so on."[103] Moreover, the telescoping of time into a present that contains both the past and future heightens the conditions for action in the unfolding now: through present sacrifice one acts not only in the present but in the past and future. The intimation that the legionnaires listening at Fiume are time travelers capable of

La nota... radiotelegrafica illustrata

UTILE E DECORO

In attesa del *disegno di legge* per la ricostruzione del campanile di Venezia, ecco un *disegno...*. pel quale il campanile di *San Marco* si trasformerebbe nel campanile di *San Marconi.*

An illustration of "San Marconi" from 1903, showing the effect of the wireless on the collective imagination. Courtesy of the Archives of the Fondazione Guglielmo Marconi, Italy.

moving in the present across time is one not easily shaken, which is in keeping with the operations of apocalyptic rhetoric and imagery generally.[104] Where D'Annunzio's Marconi speech parts ways with traditional apocalyptic procedures is in the aestheticization of action that the wireless occasions. In a footnote that deserves a book, Avital Ronell speaks of the "somewhat unmapped access roads to fascism" that have been uncovered precisely by the discussion of the aestheticization of forms.[105] Naturally, distinctions must be observed, as she rightly does, between the excitement technology exerted on fascists and its severe grafting on to conceptions of self-identity in the German case. D'Annunzio to be sure was no Nazi.[106] Yet to the degree his calls for action at Fiume derive from such an aestheticization of wireless forms of temporality, access to the past, and acoustic practices, we will want to say that the Fiume orations align the wireless and D'Annunzio's poetry with certain future aspects of fascist cultural practices. It is to this larger area of concern that I now finally turn.

Fascism at Fiume

As I pointed out in the introduction, much recent writing dedicated to Fiume reevaluates the relationship between the events at Fiume *(impresa fiumana)* and Italian fascism. Claudia Salaris is typical of many when she foregrounds the liberating cultural, political, and sexual policies adopted at Fiume, which are strictly speaking at odds with the authoritarian and oppressive milieu of fascism. In this she assumes something like Renzo de Felice's perspective on Fiume outlined in *D'Annunzio politico* and elsewhere, in which the noted historian elides any privileged connection between the two, relying on the openness of the speeches as "that genre of documents that can be read without difficulty or in a forced fashion, in diverse ways, according to the interest and sensibility of each reader." For reasons unspecified, this is not the case with Mussolini's discourses.[107] But de Felice's and Salaris's denials of fascist reverberations at Fiume seems preemptory at best and naive at worst. To see why, consider again Derrida's choice of *il duce* to mark the mystagogue who adopts apocalyptic rituals in order to seduce politically. By deploying the term in a context of apocalyptic tones, Derrida implies some structural analogy between the founder of Italian fascism and the overlord who by seeing the future initiates others into "the mysterious secret come from beyond."[108] Although he chooses to allow the associations to linger of their own accord, our discussion of D'Annunzio at Fiume permits some further thoughts on the nature of the relation among mystagogue, *comandante,* and *il duce.*[109]

The chief question to be asked is to what degree does an apocalyptic tone "program" the *impresa fiumana* and fascism's relation with its subjects?[110] It is not necessary to examine each of the thirty-five volumes of Mussolini's collected works to recognize that Mussolini repeatedly staged fascism apocalyptically in ways deeply analogous to D'Annunzio's Fiume orations. One finds the narrative of alarm linked to the *patria*'s overcoming of life and death, a rhetoric that promises to disclose the contents of a mysterious future contained in the words and the figure of Il Duce and joined to the fortunes of the *patria,* and not least a technological frame authorizing prophecy. In each case Fiume precedes if not models some of the most important features of later fascist cultural practices. I am not suggesting, however, that Mussolini was unaware of the importance of defending the *patria* before the events of Fiume. Sternhell's magisterial mapping of the topography of the defense of the *patria* and the attraction of a dictatorial regime before the march on Rome is proof enough to the contrary.[111] But there is little in Mussolini's speeches and writings before

Fiume to suggest anything like the wedding of alarmist narratives, technologically generated apocalyptic tones, and ritualistic disclosures of the future that one finds in Mussolini's post-Fiume discourses. Indeed, as one advances into the years of regime and consensus, communication media in the form of voice transmissions over wireless radio will often provide Il Duce with access to futural readings of the *patria,* the speeches claiming to know the future thanks to the intimate coupling between Il Duce and new communication media.[112] Moreover, distilling an apocalyptic tone in D'Annunzio's and Mussolini's later speeches also allows us to take issue with one of Ledeen's conclusions, namely, that "no single position, no single ideology came from D'Annunzio's style: the method remained constant, the particular theme was subject to change."[113] Our discussion of Fiume complete, we are in a position to specify the "method" deployed by D'Annunzio at Fiume as one in league with an apocalyptic rhetoric meant to produce not only alarm in its listeners but also to mobilize attraction to the holocaustal fires in a future moment of self-sacrifice.

A careful reading of D'Annunzio's "Marconi" oration underscores how wireless technology models the deterritorialized and reterritorialized perception that both the *comandante* and Il Duce utilize for their own ends. Both create conditions in which primary acoustic moments may be inscribed within different tonal registers and recontextualized according to the frame of their choosing, be they the symbols of the Church now appropriated for the visual grammar of Fiuman sacrifice or state power and sacrifice so well elaborated under fascism. When discussing the creation of the "new fascist man," Emilio Gentile like many is content with the notion of a fascist ideology to explain the process in which a solider citizen has been "trained to consider himself a mere instrument of the State, and prepared to sacrifice his life for it."[114] Yet Gentile rarely details how fascism "trains" its adherents to sacrifice themselves or examines the reinscription of acoustic perception along lines more congenial to fascist power. Some come closer: Ledeen speaks of a "melodramatic and poetic structuring of the crowd" at Fiume and by extension in fascism; Mosse intimates in an aside that "rhythm is important" when discussing the theatrical aspects of the Fiume performance; and de Felice, recognizing only one input of the feedback loop running from the crowd to D'Annunzio, argues that Fiume was nothing more for D'Annunzio than "a great and exciting adventure, conducted under the impulse of reactions and stimuli, of external suggestions received in a very personal manner."[115] This discussion of wireless practices implies a much more significant role for the heretofore ignored materialities of communication in operation at Fiume and in fascist Italy, not just as a frame for authorizing access to oracular readings of the *patria* but equally as a model for a new mode of listening in which silence is transformed into a medium of revelation and the *comandante* becomes its agent of disclosure. There is much more to say about these procedures and future work certainly ought to examine at greater length the relation between Mussolini's voice transmissions over radio and the "absorption" of myth as a living reality by the fascist masses.[116] I return to some of these same issues in my discussion of Ezra Pound's broadcasts over Radio Rome in chapter 5, but what requires our immediate attention is the role of spacing on acoustic perception in the years immediately preceding Marconi's invention. A genealogy of wireless practices *avant la lettre* provided in the next chapter will assist us in determining with greater precision the material and symbolic procedures needed to construct something like a wireless acoustic subject.

3

STATE OF THE ART: MARINETTI'S WIRELESS IMAGINATION

WRITING IN 1908, well after the lessons of psychophysics and its associated "sciences" had been learned, Agostino Gemelli, noted Italian priest and physiologist, argued that a day of reckoning for what he termed scientific, quantitative psychology was at hand. "Everyone knows that today, when speaking of scientific psychology, one signifies with these names a psychology that, having broken every link with every metaphysical conception and completely removed from philosophy . . . wants to provide us with, let us call it, a physiology of the soul." A few pages later he articulates more forcefully the soul's exit from metaphysics and its rebirth as automaton in psychophysics: "Reducing all individual difference, introspection was banned and the subject was reduced to an automaton, the more perfect the more his reaction to stimuli could be reduced to an anonymous and impersonal number." Gemelli does not doubt that measuring perceptions among the formerly human had been illadvised, but more troubling still is the blind spot of quantitative psychology. "These authors," he writes, referring to psychophysicists Wilhelm Wundt and

Gustav Theodor Fechner, "don't admit that one might be able to acquire a firm knowledge of psychic processes by directly perceiving them in ourselves."[1] By 1908, quantitative psychology was apparently no match for the recently forged tools of psychoanalysis.

Gemelli's account may seem an odd point of departure for a chapter that takes as its object the wireless, aesthetics, and their intersection in F. T. Marinetti's Futurist manifestos. But as soon as one recognizes that every technology (the positivist quantification of perceptual capacities one among others) isolates cultural practices by reference to deficiencies, the path from applied physiology to aesthetics is a short one.[2] The difference between the technology of psychophysics with its imposition of automatic functions and the wireless demand that its signal be heard and recorded becomes entirely one of degree. Both crack the whip, making patients and operators into bodies that write themselves. Tracing a genealogy of wireless hearing and writing through experimental psychology and its countless ways of measuring perception is not, however, the primary focus of this chapter. Rather, Gemelli's evacuation of the soul leads me in another direction, namely, to the outlines of a possible media history to which the wireless belongs. Such a history asks where wireless technology collaborates with other media systems, the most significant being writing and psychophysics. To pose the nature of the transmission in these terms raises again the status of the wireless but from a different vantage point. I want to argue that the wireless does duty both as a "real" gathering technology, to borrow one of Friedrich Kittler's categories, and as a symbolic registration, that is, a mode of exchanging places symbolically. In a system of connected media, the question of content gives way before the issue of incompatible data channels and differently formatted data, where determining levels of noise and information transmitted on the wireless becomes the paramount task. Formulated negatively, we need to determine what of the real does not pass in the wireless transmission.

Three texts frame my chapter. The first is a selection from Giuseppe Sergi's *Teoria fisiologica della percezione,* a voluminous tract from 1881 that spells out in magisterial detail one scientist's attempts to measure and standardize the acoustic spacing subjects require in order to register rudimentary sound combinations. Not quite as far-reaching as his German positivist counterparts, who in Gemelli's words had performed autopsies on the soul, Sergi's work has the advantage of focusing on sounds, transmitted mechanically, which could be slowed down or sped up so as to determine the precise measurements needed to create listening subjects. By examining Sergi's results, I hope to isolate the conditions necessary for what he calls "fused associations," particularly the regimen that bodies had to pass through if they were to make the sounds join up in the nervous tissue. The experiments will offer a first order of wireless transmissions to the degree they measure the meaningful spaces between sound units.

After appropriating Sergi's notion of keyed spacing and bodies trained in repetition for wireless media, I examine two Futurist manifestos and their enmeshment in wireless practices. The first, the "Technical Manifesto of Futurist Literature," is the faithful registration by one automaton, F. T. Marinetti, of a propeller's transmissions, or as Jeffrey Schnapp concisely puts it, the place where Marinetti "first felt the need to free language from the prison house of traditional syntactical form."[3] Here, Marinetti's literary inventions, "words-in-freedom" *(parole in libertà)* and a wireless imagination *(immaginazione senza fili),* inscribe the newness of the wireless in the medium of writing; they become the indices of an altered media environment in which the components of writing are uncoupled via the installation of the wireless where Marinetti's outdated lyrical I *(io lirico),* as he calls it, once resided. In the "Responses to Objections" that follow, Marinetti becomes the wireless operator of his own brain, reproducing the technical processes of quantitative psychology that so trouble Gemelli. Marinetti learns to observe his own hand writing from the position of a detached head, which in antenna-like fashion surveys the damaged limb. Imagination is transformed into (or revealed as) a dictating machine whose transmissions to the medium Marinetti have little to do with sense. That the wireless could be appropriated as a literary practice would have been old hat for Gemelli: creating the conditions for perceptual deficiencies had led to measurable standards of what constituted reading and writing twenty years earlier. Marinetti's contribution consists in providing real technological standards for poetry, toward the establishment of recording thresholds for capturing Brownian movement.[4] Two questions loom large: how does Marinetti process information and how "state-of-the-art" are his *parole in libertà* in registering the new media situation?

The second Futurist manifesto, "Destruction of Syntax—Wireless Imagination—Words-in-Freedom," provides some answers. Rather than simply revealing a more auditory Marinetti, "Destruction of Syntax" indicates the extent to which the wireless dominates the Futurist aesthetic of endless associations and of the essential, condensed word. Of particular importance will be the extension of registering from sound to the more complex data flows of the visual. In the process, a wireless imagination begins dictating sense perceptions. The resulting *parole in libertà* make no distinction among various data streams but limit them to what another medium, writing, is capable of reproducing. Afterward, as the wireless begins to increase the velocity at which the data flows are transmitted to a writing hand, it facilitates an exchange of information between man and machine. The results are spectacular. In a radically altered media environment of wirelessly transmitted sense perceptions, the wireless invites the former poet to register the world's hot zones with a wireless imagination, reshaping him into a war correspondent who covers the newly created theaters of war. The wireless alters bodies and psyches in preparation for war-time transmissions and confirms that the question of the wireless is summed up in the question of speed. By linking the wireless transmission to speed and war, I am of course following the lead of a number of critics who place Futurism "at the culminating point of an anthropology of speed and thrill that evolved over the course of two prior centuries."[5] Hence the recurring emphasis in this chapter on psychophysics, the wireless transmission, and their entanglement in a modern anthropology of speed.

The insight into wireless technology, speed, and data streams leads me in the chapter's final pages to an examination of physiological deficiencies, namely, autism, and how the simulated wireless receiver creates conditions that short-circuit traditional associations. Short-circuited transmissions return us to a consideration of the aesthetization of forms in fascism, where they become a key to reading the fascist not just as a mystified user of technology but more speculatively as an exhibit in a history of wireless information processing.

Sound Autopsies

The Velocity of Vibration

I want to begin by sketching, if not a strict genealogy of the wireless, then a significant detail in the pre-wireless era. It concerns a sound device used to measure and standardize the optimal velocity at which nonmeaningful sounds assimilate into a perceptible unit. My intent is to outline a possible media system to which the wireless belongs, some years before its official emergence. I have chosen an Italian psychophysicist of European fame, Giuseppe Sergi, though I could easily have selected any number of names forgotten to us today.[6] Sergi recommends himself above all for the chapter "Association and Contrast among Perceptions," which appears in an early opus, *Teoria fisiologica della percezione*. There he attempts to localize a perceptive phenomenon in which movement, acting on the sensory organs, produces, in the best scientific language of the day, *un eccitamento*. According to Sergi, a unit of excitement is composed of distinct and similar elements, all measurable and calculable, though difficult to measure given the speed with which the succession of *eccitamenti* occurs. He gives himself the following task: "But the speed is so large in the sequence that a unit of movement is combined with the one that follows and so on until the end of the action is reached. Where does this fusion take place?"[7] The answer lies in "the nervous substance," a material that apparently requires a certain accumulated force if the smallest unit of *eccitamento* is to be registered. The key to a successful registering of *eccitamenti* consists "in the speed with which they follow each other, so as to fuse with a larger unit, which is a composed unit, and which therefore appear as elemental."[8]

To prove his thesis, Sergi conducts a series of auditory experiments on a number of patients.

> The excitation of hearing provides us with a clear example of sound vibrations, from which a sensation is induced. If more than a sixteenth of a second passes between a vibration and another, there is no fusion, and the sequence is ineffective in creating a sense effect. The nervous tissue cannot retain for even one vibration the excitation that it receives, for it is an extended period of time, i.e., a sixteenth of a second. The second vibration that arrives doesn't cohere with the first, which is gone without a trace. Lessening the distance between one and the other brings on fusion.[9]

What is Sergi measuring in the passage? Clearly, he is focusing on the amount of time between "sound vibrations" that an individual requires for the vibration to be "retained" *(conservato),* that is, stored by the nervous tissue. If more than a certain unit of time passes, a sixteenth of a second by Sergi's measurements, then the nervous tissue is incapable of storing the excitation *(eccitamento)* that has been transmitted to it. As a first response, Sergi is measuring the silence necessary for one sound to be joined to another in what he defines soon after as "a composed unit for distinct perception."[10] Let's call this required millisecond "keyed silence" to mark its mechanically manufactured status. But necessary for what one wants to know. Not lacking a certain romantic touch, Sergi finds his answer in the fusion that occurs between vibrations when the distance between them has been reduced to a level that allows the nervous tissue to store the transmitted excitation.

Leaving aside the inexactitudes of the experiment and its nineteenth-century excitable nervous tissue, Sergi's tests essentially quantify the length of time required for sound vibrations to amass into a complex unit of perception. Anything greater than a sixteenth of a second and the second vibration will find itself without its better half. Although his emphasis is on reducing the distance between sounds by increasing the velocity between them, there must, by his own definition, exist a minimal, keyed-in unit of spacing—silence given the auditory nature of the experiments—for a fusion to be registered. To a remarkable degree then, Sergi's formulation demonstrates a fundamental tenet of media theory. The space between vibrations, even when shortened to meet his requirements, must always be present if a meaningful transmission of sound may be said to have taken place. Anything that falls outside of the allotted space is "gone without a trace." Later, Sergi will begin piecing together the various sense media into a model of association, and there as well he listens to the background of an omnipresent spacing on which the vibrations are imposed. Implicit in the model is the early recognition of what is often considered of contemporary provenance: noise as a condition of meaning or, translated into wireless terms, static as the sine qua non of the transmission. A precondition for something to be written down, picked up, or perceived as a meaningful vibration (which amounts to the same thing) is in fact a relation to its background or static: letters and sound vibrations are determined by the space between them.[11] Nowhere is that more clearly demonstrated than in the wireless combination of dots and dashes for letters and static for noise.

Sergi goes on to isolate a second condition for the possibility of fused associations, namely, bodies trained in the art of repetition:

> The phenomenon of association, as much in this simple form as in those more complex, admits of another condition, which is tied to the repetition of actions, or reproduction. If every time an action were to be repeated, appearing as new, the repetition would not have any effect. If instead the past is repeated, one finds a recognition of this, and the repetition confirms the first.[12]

The passage reads like a regimen bodies must pass through if they are to muster the milliseconds needed to make the vibrations link up in the body's nervous tissue. Not only randomness must be reduced but also whim and forgetfulness, because only by not forgetting can the body repeat and renew *(rinnova)* the past. Again it bears remarking that Sergi's bodies are not agents in any strict sense of the word but rather an amalgam of nervous tissue whose thresholds of sound registration are measured and then trained to reproduce recognition. Sergi's test localizes reaction thresholds, establishing a point of connection between technology and the body. If a certain velocity is not maintained between vibrations, then those on the end of these transmissions are not considered listeners in the traditional sense. When the body has learned an order of association by repetition and the optimum velocity for transmissions has been reached, only then may we begin to speak of something approaching agency, subject and actor. Even so, it seems an inappropriate descriptor for the final result since by the experiment's end what is finally measured is a body's response time to noise. David Wellbery's insight into corporeality and technology echoes this thesis: "The body is not first and foremost an agent or actor, and in order to become one it must suffer a restriction of its possibilities: the attribution of agency is a reduction of complexity."[13] Sergi's noise-producing machine commands bodies to form meaningful sound units by standardizing a velocity that will allow bodies to associate one sound to another. Sergi distills his findings regarding sound association and the mechanics of reproducing configurations into a theorem: "Association of sounds presupposes a fact, the act of reproduction, which we saw presupposes for its part the association of sounds."[14]

Nervous Tissue, Sound Vibrations

While I do not want to overstate the importance of Sergi's experiments for a tentative media history of the wireless, it seems clear that the wireless and Sergi's infernal sound machine share a number of traits that link them in a network of early wireless practices. First, there is the body of the experimental subject as a surface of nervous tissue to which Sergi transmits in order to determine the optimal velocity at which his hearers put together one sound sample with its successor. Translated into the terms with which I began this chapter, Sergi's experiments operate symbolically to the extent the body is inserted into a structure of meaningful, spaced intervals. As it learns by repetition to associate sounds, the body comes to resemble the listener known so well to us today. Similarly, the wireless transmits to bodies, but rather than simply making bodies into adept listeners, it creates wireless operators that are capable of picking up and relaying messages. Although no one speaks of the nervous tissue of the *marconisti,* they were also able to associate sounds into perceptible units that could then be transposed to paper via their writing hands. The distinction between the two lies in the wireless operator's much smaller noise-to-signal ratio and in the limited register in which he hears: only dots and dashes qualify as a wireless message.

Second, both Marconi's wireless and Sergi's sound machine are sources of vibrations. But where Marconi's invention vibrated with staccato bursts of energy, Sergi's is rather a mechanism for transmitting random configurations that measure the point at which an experimental subject is capable of hearing meaningful sounds. The sound machine is not directed toward producing meaning or truth but instead generates noise, chopping up sound into discrete units and then measuring bodies in terms of forced intervals. This point is easily lost in Sergi's summation, where he makes no mention of any vibration-producing keyboard or the soundless space bar that surely accompanied it. One wonders as well whether at any point after these initial experiments Sergi utilized a gramophone, which would have offered the possibility of a further standardization of sound intervals. Sadly, Sergi makes no mention of it in his later works. As it is, these mechanically produced sounds are not human, not speech at all, but unarticulated samples that of themselves have no meaning. They come together against the background of a ubiquitous noise that by Sergi's own definition would simply be sounds that lack sufficient *allegro* spacing to inscribe themselves in the body's nervous tissue.

The wireless also isolates sound units against a background of noise, standardizing differences among the various operators, so that a practice of distraction, defined as not listening to the noise but to meaningful signals, was developed, first by Marconi and later institutionalized by the Marconi Wireless Company in its many wireless training academies. We already know that Marconi's primitive transatlantic wireless was in fact a one-way telephonic headset capable of tuning into the noise on the ether, out of which Marconi's ears could distinguish or hallucinate a meaningful "S." Future versions of the Marconi wireless certainly improved on the prototype by increasing the transmission distance, but just as key was the standardization of signals, of what qualified as a dot or dash that could be picked up with a minimal amount of patching up. Limiting the complexity of what the operator heard was necessary if the dots and dashes were to be transposed to paper, and this became more so with the proliferation of wireless shorthand codes that sought to standardize what was sensical and not on the wireless.[15] Both Sergi's experiments and the wireless practices that developed in the first decade of this century attempted to set standards for associating sounds correctly, but the wireless joins what are in themselves meaningless sound configurations with alphabetized bodies learned in the art of associating not simply individual sounds, one with the other, but these same sounds with their twenty-six-letter counterparts (twenty-seven if we include the keyed space).

In other words, the Marconi wireless transmits not with a voice but writes with the alphabet the sounds its operators have been trained to hear. What their ears disregarded was the debris of the transmission, the nonsense dots and dashes that lacked an association with the sound bits that followed or preceded

Marconisti *in training at a "wireless" academy. Reprinted by permission.*

them. Sergi's experiments indicate the precise level at which a sound becomes meaningful; it is a machine that could dictate nonsense to ears not yet versed in meaning or spacing. Similarly, the wireless draws up a relation among spacing, sound, and meaning and utilizes the hands available to it for writing. Consider the unimaginability of the nonassociated sounds Sergi's machine transmits and the insufficient spacing that make a hermeneutic decoding beside the point: a noise-transmitting machine that operates below its specs promotes no medial selection. The wireless represents an engineering advance on such a machine since it encodes the real; indeed, its purpose is to relate discrete elements by providing a key, the Morse code (and soon after shorthand wireless codes), with which to associate sound elements. The wireless operator is no superfluous stand-in for the wireless nor is he a simple extension of the technology but one able to transpose the static of a transmission into a model of meaning via his alphabetized, writing hand. The wireless selects what the semi-listening operator will write by requiring him to hear the proper spacing between sounds, a process one step removed from Sergi's machine. The primary difference between Sergi's and Marconi's machines may be found in the wireless's command to limit the noise of nonkeyed spacing to what another medium, writing, could register.

The Wireless Imagination

Instructions for Use

Sergi's writings are useful for unearthing the media forebears of the wireless and for offering a useful comparison with the enhancements that followed. They also demonstrate that the symbolic function of his machine as opposed to its "real" ability to capture the immeasurable static of nonmeaningful sounds is limited. Consequently, he fails to describe fully the physiological effects of the machine on the individual body; indeed, in the five hundred or so pages of quantifying, standardizing, and torturing individual ears, nowhere does Sergi name one subject: individual cases are erased in his search for the measured and cataloged human. This is symptomatic of the disappearance of the individual in a psychophysics that quantifies automatic functions. Of great interest would be an account that chronicles the symbolic features of the wireless with all its corporeal effects, revealing where the reproduction of wireless practices mirrors physiological ones. These would include not only the traditional grab bag of agraphia (pathological loss of the ability to write) and alexia (loss of the ability to read) but also of autism. In the "Technical Manifesto of Futurist Literature" and "Destruction of Syntax—Wireless Imagination—Words-in-Freedom," Marinetti details minutely what happens when a poet attempts to

simulate wireless practices on his own skin; he engages in a bodily reproduction
of technical processes that teach him how to hear and what to observe and in
the process detach his head from a writing hand. The resulting "literature" is
revealed as a transposition of media and an assault on the notion of sound as a
complex of data that is impossible to put into writing, while Marinetti, the poet,
becomes a wireless operator transposing the noise of modern life into writing.

In the following sections, I want to comment on the 1912 "Technical Manifesto" as a site in which writing, sound experiments, and the wireless meet. As
is the case whenever media connect, there are incompatible data channels, the
differently formatted data of the wireless and the book. What consequences
ensue when a writer simulates wireless reception of different streams of data?
Marinetti's simulation of a wireless receiver in his brain will underscore the
wireless's speed and success in detaching limbs from the body and indicate a
further function for the device: a sensory deprivation or physiological handicap in the way the wireless earmarks signals for reception and a certain mode
of elaboration. Aesthetic strategies will be disclosed as the means by which a
medium, in this case writing, registers Brownian movements via a simulation
of wireless functions.

The manifesto opens famously with a scene of dictation between a biplane
propeller and Marinetti, who, sitting atop the gas cylinder, tunes his hearing
to the sounds of the propeller behind him. What he writes at that moment and
what follows, he tells us before proceeding to his bolded letters and mathematical signs, is a transposition of the sounds of the propeller into the medium of
writing: "This is what the whirling propeller told me, when I flew two hundred
meters above the mighty chimney pots of Milan. And the propeller added."[16]
Marinetti takes dictation from the propeller and so puts into practice one of the
key Futurist tenets of the manifesto, as outlined in paragraph 11: "To listen to
motors and to reproduce their conversations."[17] That a propeller's dictation is
a condition for the production of the manifesto points to a first level of media
connectivity between Marinetti's ear and the transmitting propeller. Where
one might have expected the usual paeans to Marinetti's *genio,* which are not
missing in the manifesto, there is a sound-producing machine that dictates not
only how it is to be heard but how it is to be written down. Marinetti becomes a
medium for the propeller transmissions and begins bracketing real data flows.
The result, at least in part as Jeffrey Schnapp has pointed out, is a symbolic selection from the propeller's real sounds. What requires comment is how Marinetti
processes this particular data flow and whether I am justified in joining this
processing, once called literary production, to wireless media.[18]

Even though the propeller's transmissions begin *in media res,* marking ad-
jectives and adverbs for cancellation, Marinetti's strategies for simulating the
wireless media's effects on aesthetic production are to be found elsewhere. Two

concepts, or more precisely, a condition for production and a mode of association, delimit the engagement between propeller and medium. The first, wireless
imagination *(immaginazione senza fili)*, is an obvious reference to the Marconi
wireless, the telegraph without wires *(telegrafia senza fili)*, though Marconi, another wireless medium, nowhere receives credit. The second, words-in-freedom
(parole in libertà), registers in script the dictations sound machines make to an
imagination without wires that is connected, wirelessly, to unconscious writing
hands. To help make clearer the nature of these connections between writing,
sound, and the hand, I want to delimit first the nature of Marinetti's simulated
wireless imagination. Not surprisingly, some of the same ghosts that haunted
D'Annunzio and the casts of chapters 1 and 2 make a curtain call here.

The section on the wireless imagination appears in the densest section of the
manifesto and only after Marinetti has established the conditions for transposing noise into writing. He first proposes a new psychology of matter, one that
coincides with what has been revealed to him high above in the biplane:

> The man sidetracked by the library and the museum, subjected to a logic and
> wisdom of fear, is of absolutely no interest. We must therefore drive him from
> literature and finally put matter in his place, matter whose essence must be
> grasped by strokes of intuition, the kind of thing that the physicists and the
> chemists can never do.
>
> To capture the breath, the sensibility, and the instincts of metals, stones,
> wood, and so on, through the medium of free objects and whimsical motors.
> To substitute for human psychology, now exhausted, the lyric obsession with
> matter.[19]

The passage differentiates starkly between "literature" and that which does not
depend on logic or an author's erudition for its registration. The latter can be got
at in isolated episodes of intuition, when objects *(in libertà)* may be caught via a
process of *sorprendere*. The word deserves our attention for it appears whenever
Marinetti is instructing his readers on how to engage the movement of objects.
Sorprendere is a composite of *sor-* and *prendere*, literally *prendere dal di sopra*
(to take from above), which in turn has come to mean "to capture suddenly,
especially when attempting something dishonest."[20] The reference to Marinetti's
vantage point, high above Milan in a plane, is clear enough, but it also points
us in the direction of the kind of listening (and recording) that Marinetti is at
that moment himself practicing. Listening to the propeller, Marinetti captures
its respiration, its sensibility and its instincts, as he later puts it, not by a direct
transposition of the sounds but rather through a process of *sorprendere*. What
form this and the "jerks of intuition" *(colpi d'intuizione)* may take remains unclear, although he indicates that listening to "free objects" *(oggetti in libertà)*
and subsequently transposing them into writing will rid us of literature. Where

there was before only a rotten human psychology, now there will be a lyrical obsession of material.

Let me be clear. I do not believe that the death of literature at the hands of the wireless is simply another Futurist conceit, which would come as no surprise considering Marinetti's flare for exaggeration. Marinetti's lyrical obsession with matter demonstrates rather that with the growth of wireless media, knowledge, aesthetics, and literature count for less. "Look not to give human sentiments to matter," he writes on the next page. In its place, information comes to the fore as that which makes a difference, information about matter, "its different governing impulses, its forces of compression, dilation, cohesion, and disaggregation, its crowds of massed molecules and whirling electrons."[21] Marinetti has no need for literature since it fails to record adequately the movements of matter. He requires sickness and the dissociation of limbs that limit transposition to a dictation that merely registers matter. Thus, there is no need to call in an author. The manifesto echoes the results of Sergi's sound experiments on association in which unnamed human subjects disappear in information concerning their automatic capabilities. Where those experiments succeeded in identifying the precise sound interval necessary to create a listening agent, Marinetti invites us to refrain from humanizing matter; he directs our attention to the possibility that only another unit of matter, the formerly listening literary man, is capable of registering the Brownian movement of fellow matter. To call the process literature occludes the place the wireless occupies in a new media environment in which only information matters.

One further note: Marinetti's registration of nonhumanized matter moves through the medium of poetry, which will be the symbolic writing down of sounds whose transmission never ends. No spirit stands behind Marinetti's lyric of matter; instead, poetry functions as a specific carrier of information. This accounts for the sheer randomness that governs syntax, the dispersal of nouns as they arise, and Marinetti's strategy outlined in the manifesto's opening paragraph: "One must destroy syntax and scatter one's nouns at random, just as they are born."[22] Marinetti's instructions to the Futurist *in potentia* are a command to write down in poetry the incessant sounds that matter makes, recording wireless minutiae, the data flow of objects that sound in the real. That the sounds never cease or that the data flow of *oggetti in libertà* and the medium of poetry may be incompatible are issues we will return to momentarily. In the meantime, note again the conditions Marinetti sets for transposing noise into writing. He equates visual and audio sense media under a rubric of wireless transmissions in which objects dictate. A body writes down the endlessness of movement, with free objects functioning as Sergi machines that operate below the threshold of meaningful combinations of sound units.

Simulating the Wireless

Inserted between the sounds of modernity, the propellers and automobiles that dot the manifesto's landscape and the body that registers their movement is the imagination without wires. That the term resonates today, eighty years after its introduction, is a testament to Marinetti's extraordinary grasp of a publicist's vocabulary.[23] Unfortunately, the nebulous nature of the concept and the short space Marinetti devotes to it in a manifesto dedicated to technology and literature require that we proceed carefully. The reference to wireless imagination appears only once and then in no seeming relation to what follows:

> Together we will invent what I call the *imagination without strings [l'immaginazione senza fili]*. Someday we will achieve a yet more essential art, when we dare to suppress the first terms of our analogies and render no more than an uninterrupted sequence of second terms. To achieve this we must renounce being understood. It is not necessary to be understood. Moreover we did without it when we were expressing fragments of the Futurist sensibility by means of traditional and intellective syntax.[24]

Marinetti's invention of the wireless imagination is an odd formulation, especially considering what we know of the two terms. On the one hand, the term "wireless" seems natural enough in a paragraph that concerns aesthetic considerations and the more essential nature of accelerated transmissions. Landini's boyhood spent vandalizing telegraph and telephone wires and Marconi's destruction of heavy matter at the Villa Marconi with the first wireless transmission indicated not only a distrust of wires but also formalized the device's relation to the more connected and hence more real connections of mental telepathy: to be precise, wires weigh down lines of communication. Marinetti echoes these considerations in the manifesto, equating greater distance between analogies with more solid relationships between them:

> In this there is *an ever-vaster gradation of analogies,* there are ever-deeper and more solid affinities, however remote. Analogy is nothing more than the deep love that assembles distant, seemingly diverse and hostile things. An orchestral style, at once polychromatic, polyphonic, and polymorphous, can embrace the life of matter only by means of the most extensive analogies.[25]

On the other hand, Marinetti is taking dictation from a propeller, the fragments of which appear in their traditional syntactic accoutrements. He repeatedly urges his readers to "introduce three elements into literature that have been neglected," the first of which is noise *(rumore)*. The difficulty of the formulation, it seems to me, lies in speaking of imagination, a term whose origins and celebration are located in Romanticism, in the same vein as dictation and writing down sounds. Two questions follow: who or what dictates when a wireless imagination is engaged in registering the new sounds of modern life, and what would a nondictated and hence purely inspired cultural product of imagination look like?

In answer to the first, Marinetti is uncertain; he hedges his bets by coupling the creative spirit with a lucid will. "Every creative spirit can confirm during the work of creation that intuitive phenomena combine themselves with logical intelligence. It is therefore impossible to determine exactly the moment in which unconscious inspiration ends and lucid will begins. Every now and then this latter one suddenly generates the inspiration, other times rather it accompanies it."[26] Inspiration arises unconsciously with a marked lucidity, or intuition produces a hallucination that is the first act of poetic production. Afterward, a pen marks down in words what imagination has given birth to. Such a formulation recalls poetry's role within a conception of the Romantic imagination. Writing in the context of Romanticism and the discourse network of 1800, Kittler equates imagination with poetry: "It is precisely the translation of other arts into a nonmaterial and universal medium that constitutes poetry. This medium is variously labeled fantasy or imagination. Imagination generically defines all the arts, but it specifically defines one highest art. Only poetry can claim 'the imagination itself, that universal foundation of all the particular art forms and the individual arts' as its proper material."[27] On first reading, we might also say that poetry claims Marinetti's notion of imagination as its proper material, for the manifestos never refer explicitly to writing as a medium. Interestingly, Marinetti joins the larger realm of the creative spirit, itself an unrecognized dictating machine of poetry with intuition and wireless imagination. A century before, poetry was both the primary medium for imagination and the generator of flights of fancy that were written down. Marinetti essentially substitutes a hallucinating creative spirit that he calls a wireless imagination for a previously hallucinating Romantic one, and so exposes the field of poetry and imagination that Kittler describes so acutely. With the introduction of an imagination made wireless, Marinetti gestures toward a radically different form of literary production based on the dictation of *oggetti in libertà* to a writing hand. The premise is that poetry must now reproduce and transpose a propeller's movements.

Yet a dictation is a dictation whether the sender be a universal imagination or recently invented wireless one. Nowhere is this more apparent than in the wireless registration of sense data, the *parole in libertà*. The term first appears near the end of the manifesto, though Marinetti, as is his wont, has previously revealed its major component:

8. There are no categories of images, noble or gross or vulgar, eccentric or natural. The intuition that grasps them has no preference or *partis pris*. Therefore the analogical style is absolute master of all matter and intense life.

9. To render the successive motions of an object, one must render the *chain of analogies* that it evokes, each condensed and concentrated into one essential word.[28]

Marinetti sets up a duality between "categories of images" and "chains of analogies," two seemingly different registers with which intuition maintains a relationship. For its part, intuition both perceives the successive movements of an object in motion, occupying the role the senses have in Sergi's experiments, and necessarily gives *(bisogna dare)* the chain of analogies that the object evokes. A previously encoded version of the real, a visual category with the reference to images, gives way before an auditory process based on a calling forth *(evoca)* that is finalized in another symbolic encoding, more condensed and essential than its predecessor. Intuition dictates in both cases, but only in the latter does the registration more closely correspond *(analogo* from the Greek) to the real. In the widening space that he grants his *parole in libertà,* Marinetti moves decidedly away from a Romantic conception of imagination to one more properly resembling a twentieth-century wireless machine.

The consequences of reading the visual as just another data stream are significant. It forces us to reconsider enlisting Marinetti in the ranks of the denigrators of vision most recently described by Martin Jay or associating him with a more auditory strand of modernism as many of Marinetti's Italian commentators have done.[29] In my view, both miss the mark for a number of reasons. First, affiliating Marinetti with the less visual may help account for the genesis of the *parole in libertà* but ignores the highly visual content of Futurist-shaped writing, the free-word poet Francesco Cangiullo's auto-illustration *Fumare* being the classic example.[30] Second, neither Jay nor Marinetti's Italian critics locate, in a sustained or systematic way, the cause of Marinetti's heightened auditory sensibility in a simulation of wireless practices, nor do they successfully link intuition to dictation or the endless association of condensed words to sense transmissions. It is not that Marinetti is somehow less visual or more auditory but rather that both become mere data flows in a system in which a wireless imagination (call it intuition or the creative spirit) dictate the sense perceptions of objects that sound in the real.[31]

The following year's manifesto, "Destruction of Syntax," bears this out. There Marinetti imagines a friend, either lyrically gifted or traumatized, who recounts a particularly intense experience:

Now suppose this friend of yours gifted with this faculty finds himself in a zone of intense life (revolution, war, shipwreck, earthquake, and so on) and starts right away to tell you his impressions. Do you know what this lyric, excited friend of yours will instinctively do? He will begin by brutally destroying the syntax of his speech. He wastes no time in building sentences. Punctuation and the right adjective mean nothing to him. He will despise subtleties and nuances of language. Breathlessly he will assault your nerves with visual, auditory, olfactory sensations, just as they come to him. The rush of steam-emotion will burst the sentence's steam-pipe, the valves of punctuation, and the adjectival clamp. Fistfuls of essential words in no conventional order. Sole preoccupation of the narrator, to render every vibration of his being.[32]

The simulation of wireless practices moves through the poet who, unlike his Romantic predecessor, is given no time to reread and edit what was formerly inspired by imagination or the universal medium of poetry. Marinetti will not waste time *(non perderà tempo)* with the deliberate pace of editing and re-examination; only the speed of the transmission will determine its merits as lyrical production. The dictation of sense perceptions by a wireless imagination, the *parole in libertà,* makes no distinction among various flows of information. Each is transposed (and limited) to what another medium, writing, is capable of reproducing. This explains the missing position of the tragically isolated poet in the manifestos, who would have been left alone in the previous century, inebriated with his Muse. In a wireless world, the poet is commissioned to simulate the wireless operator, not simply by picking up and registering what is being transmitted on an auditory channel, but by associating all sense perceptions picked up by the wireless imagination with their written equivalent. He searches out the "intense zone of life (revolution, war, shipwreck, earthquake, etc.)" both for the sensory excitements that he will register and for its impact in loosening the ties that formerly bound him to rewriting and editing.[33] Wireless dictation allows no time for either as they merely serve to distance the real from its symbolic encoding. A second theorem of the wireless transmission might be summarized this way: the success of a transmission will ultimately be judged by its speed.[34]

The War Correspondent

The wireless incessantly demands bodies that can hallucinate a meaning out of its dots and dashes, expanding its domain to include wartime transmissions. D'Annunzio's revealing remark that both wireless and poetry were lethal instruments of war confirmed this, but where one found the wireless programming *marconisti* to hear the distress call, in Marinetti's case the wireless demands more, increasing the scope of what qualifies as a signal from the auditory to all the data streams a wireless imagination can exhaust in script.[35] *Parole in libertà* become the indices of the growing powers of wireless technology, which is revealed as a transporter of information.

A war covered by the wireless is no longer war but a theater of war, "and its sea of corpses only simulacra behind the screen of which various technologies fight for their or our future."[36] The poet who goes to war armed with his wireless imagination is a made-over medium for *parole in libertà,* a wireless operator or, better still, a war correspondent because only war correspondents are trained to cover theaters of war:

> And in order to render the true worth and dimensions of his lived life, he will cast immense nets of analogy across the world. In this way he will reveal the analogical foundation of life, telegraphically, with the same economical speed that the telegraph imposes on reporters and war correspondents in their swift reportings. This urgent laconism answers not only to the laws of speed that govern us but also to the rapport of centuries between poet and audience. Between poet and audience, in fact, the same rapport of centuries exists as between old friends. They can make themselves understood with half a word, a gesture, a glance.[37]

Marinetti knew the difference between poet and war correspondent intimately, having successfully served as a practitioner of the new genre in Libya and Bulgaria. Indeed, Marinetti refers to his Libyan tour of duty in the same manifesto: *"as Arabs who looked indifferently at the first airplanes in the skies of Tripoli."*[38] In a letter of the same year to Aldo Palazzeschi, he describes more fully his experiences: "I left Tripoli with deep-felt sadness, since I spent the two most beautiful months of my life there. I helped and participated in that which was the most violent, the most virile and the most heroic."[39] The result was *Battaglia Peso + Odore, parole in libertà* wirelessed from the Libyan conflict, and *Zang Tumb Tuuum,* the registration of artillery barrages in the assault on Adrianopolis. Besides traveling papers, what does a young man aspiring to war correspondent require? First, he must be able to turn himself into a moving listening post in order to register the *fondo analogico della vita,* best translated as the corresponding basis of life in written form. He will do so by rigorously imposing a telegraphic economy of words on himself, what one English author ten years before had described as style in the age of machinery, and what Marinetti calls the laws of velocity that govern every transmission.[40] Acoustical, visual, and olfactory sense media become flows of information to an imagination made wireless.

Many examples could be cited in order to show just how closely identified the war correspondent is with a sense-registering wireless imagination.[41] But

Michael Herr's account of his experiences in Vietnam lends itself particularly well to a comparative reading with the manifesto, despite its appearance some sixty years later. Few are as honest as Herr in pinpointing not only the thrill of war but the process whereby the war correspondent forges and converts these stimuli into a shell that can "thrive" on such shocks, to borrow Hal Foster's terminology. It is his first day on the job, and Herr is reeling from a wireless imagination that is unflagging in its data recognition:

> It was like a walk through a colony of stroke victims, a thousand men on
> a cold rainy airfield after too much of something I'd never really know, "a
> way you'll never be," dirt and blood and torn fatigues, eyes that poured out
> a steady charge of wasted horror. . . . I couldn't look at anyone more than a
> second, I didn't want to be caught listening, some war correspondent. I didn't
> know what to say or do, I didn't like it already. When the rain stopped and
> the ponchos came off there was a smell that I thought was going to make
> me sick: rot, sump, tannery, open grave, dumpfire—awful, you'd walk into
> pockets of Old Spice that made it even worse.[42]

The move from visual acquisition ("eyes that poured out a steady charge of wasted horror") to the auditory ("I didn't want to be caught listening") to the olfactory *parole in libertà* (rot, sump, tannery, open grave) all meet up in Herr's war correspondent. With the senses doing overtime for a registering consciousness, the greatest danger is clearly sensory overload, the possibility that the zone of intensity will become too hot for a dictating imagination: "Overload was such a real danger, not as obvious as shrapnel or blunt like a 2,000-foot drop, maybe it couldn't kill you or smash you, but it could bend your aerial for you and land you on your hip. Levels of information were levels of dread, once it's out it won't go back in, you just can't blink it away or run the film backward out of consciousness."[43] Herr's account surely reflects an advanced state of technological dread, with film more or less occupying the role of the wireless in the manifestos. "Life-as-movie, war-as-(war) movie, war-as-life," he writes at one point.[44] Marinetti too shares Herr's levels of dread, despite his best attempts at inoculation. The frequency with which Marinetti qualifies his initial statements as being neither categorical nor absolute is one small indication; the lamentation that he is "forced, in fact, to use everything to be able to show my conception" is another, which rings false considering the obvious relish Marinetti takes in the exposition of his thought. What keeps Marinetti and Herr from information overload is precisely the gap between the real and its symbolic encoding, the gray area in which a wireless imagination gathers information.

Consider again the manifesto "Destruction of Syntax":

> When I speak of destroying the canals of syntax, I am neither categorical nor
> systematic. Traces of conventional syntax and even of true logical sentences

will be found here and there in the words-in-freedom of my unchained lyricism. This inequality in conciseness and freedom is natural and inevitable. Since poetry is in truth only a superior, more concentrated and intense life than what we live from day to day, like the latter is composed of hyper-alive elements and moribund elements.[45]

Marinetti celebrates the ease with which the reproduced wireless practices will rid the poet of syntax. Note, however, that their traces will not disappear entirely in the new wireless world. They remain to the extent the channels (canali) on which syntax transmits cannot be completely shut down; they are inevitable in a context of media transposition from sense data to the writing hand. It may be that the paramount issue has less to do with measuring levels of dread in the manifestos than with how writing is forced to catch up when the primary concern of a new media situation becomes one of technological efficiency.[46] What appears in the manifestos as trepidation in front of the wireless Moloch or more generally as an aesthetic strategy of incorporating wireless practices, installing receivers where brains once existed, serves only to mask the competing technologies of writing and the wireless. In the competition that ensues, there is the distinct possibility that data transmitted in the hot zones, more commonly referred to as assaults on the senses, may in fact be incompatible with writing. Herr calls it overload, the informational death of a thousand cuts that awaits the writing war correspondent. I want to ask how much of wireless simulation and its attendant consequences on the body are based on an experiencing agent able to associate sounds (Sergi) and more advanced data flows (Marinetti), and how much on a nonexperiential wireless technology. Herr's account helps draw the distinction between Sergi's sound machine and Marinetti's wireless imagination in the increased channel capacity with which Marinetti equips the latter by expanding the associations from the auditory realm to the visual and olfactory. To increase the capacity of a system that could send only a limited number of signals per minute, i.e., Sergi's system of sounds, Marinetti essentially increases the information content of each transmission by designing *parole in libertà* that can optimize the recognition capabilities of the recently installed and simulated wireless receiver.[47]

Before examining the blueprints of a simulated wireless receiver, it is important to note that checking for heightened information in the manifestos offers alternate readings of Futurism. For instance, if the destruction of the lyrical I in Marinetti is the result of the wireless forcing another medium, writing, to sink or swim, that is, to give up the less efficient author and his hermeneutic commentators, then the question of dread may simply be beside the point. The "Technical Manifesto" and "Destruction of Syntax" will simply mark the ascendancy of the wireless over agent and aesthetics and reveal them to be formulations from a previous age when writing poetry was the universal medium of imagination. Additionally, a focus on information and the speed of transmission will successfully account for war's continuous appearance in the manifestos, reminding us how frequently wireless technology operates in conjunction with war, combining and transmitting a command "composed of velocity and information."[48] The solidarity of war, wireless technology, and poetry was evident in D'Annunzio, particularly when he utilized the Marconi relay at Fiume to replay compulsively all of Italy's humiliations. In Marinetti, simulated wireless practices force a writing hand to register sense transmissions instantly from a wireless imagination. As the distance between *oggetti in libertà* and the registering apparatus shrinks in the search for optimum transmission speeds, spontaneous and unconscious bouts of wireless execution become more common. That such moments are coextensive with war in distributing bodies according to their transmission speeds is a possibility I entertain in the following section.

Entropy and the Wireless Transmission

How exactly does a wireless imagination work? For all the fanfare surrounding it and the *parole in libertà* in the "Technical Manifesto," Marinetti is short on the details. Only in the "Answers to Objections" do we discover how wireless practices detach heads and limbs:

> After many painful and unrelenting hours of work, the creative spirit is suddenly free of the weight of all obstacles, and becomes, in some way, the prey of a strange spontaneity of conception and execution. The hand that writes seems to detach itself from the body and freely stretches itself from the brain, which also in some way has been detached from the body, becoming air-borne. It looks down on the unexpected phrases that take leave of the pen with an awesome lucidity.[49]

The wireless uncouples the subroutines of writing by forcing the creative spirit, what Marinetti displaces metonymically with intuition or imagination, out into the open as a dictating, sense-registering machine. After several difficult hours of "unrelenting work," Marinetti's spirit tires of swimming against the current and moves in a more congenial direction, where it becomes prey to a strangely spontaneous combination of conception and execution. A relation emerges between the writing hand cut off from a controlling consciousness, a detached brain armed with visual recognition capacities ("looks down from above with an awesome lucidity") and an unmentioned third term, the missing, dictating wireless imagination. The suggestion of a circuitous relation and the ease with which I have associated the wireless with a machine in this study point us in the direction of the second law of thermodynamics for an explanation of how a simulated wireless works. The law states that in any interaction in which energy is exchanged, the amount of thermal energy convertible to mechanical energy is diminished. In such interactions, entropy—a measure of unusable energy—increases.[50] In other words, when you do a job, a part of the energy is expended as loss or entropy. In the 1950s, a number of theorists, most notably Claude E. Shannon, extended entropy from its thermodynamic manifestation to what we know today as modern information theory. In Shannon's classic definition, entropy measures "how much information may be communicated and with what degree of fidelity, within certain constraints such as the amount of energy available in the communications system and the strength of any intrusive undesired signal (noise)."[51]

One of the first to recognize the importance of Shannon's work was Jacques Lacan. Writing within the context of Freud's pleasure principle, he makes clear the connection between entropy and the machine:

> Mathematicians qualified to handle these symbols locate information as that which moves in the opposite direction to entropy. When people had become acquainted with thermodynamics, and asked themselves how their machine was going to pay for itself, they left themselves out. They regarded the machine as the master regards the slave—the machine is there, somewhere else, and it works. They were forgetting only one thing, that it was they who had signed the order form.[52]

Of course, no textual justification exists for labeling Marinetti as the unconscious master of the machine described by Lacan; if anything, the manifestos continually celebrate signing off on the order form. For instance, consider the following from "Multiplied Man and the Reign of the Machine": "Hence we must prepare for the imminent, inevitable identification of man with motor, facilitating and perfecting a constant interchange of intuition, rhythm, instinct, and metallic discipline of which the majority are wholly ignorant, which is guessed at by the most lucid spirits."[53] In the ever shortening distance between man and machine, the wireless serves as a means for facilitating the incessant exchange of intuition and rhythm between man and motor (airplane, automobile) that is finalized in the "inevitable identification" of both. This is no medium arising somehow out of human desires or experience to justify the thesis on prosthesis known so well to us today but technical media working together so as to devise better strategies for capturing the real.[54] The wireless imagination registers real sounds and then translates them into their phonemic, alphabetic equivalencies. After Marinetti begins simulating a wireless receiver in his brain, there will be another source of information concerning the transmission standards of the human, an engineering advance on Sergi's experiments of thirty years prior that quantified meaningful sound associations.

But let's return to entropy and the wireless. Once one recognizes that a circuit exists between the machine and its user, one finds oneself immediately in the realm of energy.

> Now, this fact turns out to have a considerable importance in the domain of energy. Because if information is introduced into the circuit of the degradation of energy, it can perform miracles. If Maxwell's demon can stop the atoms which move too slowly, and keep only those which have a tendency to be a little on the frantic side, he will cause the general level of the energy to rise again, and will do, using what would have degraded into heat, work equivalent to that which would have been lost.[55]

The passage requires elaboration. Where a circuit runs between machine and user, it will by necessity show a degradation of energy, a loss, what Lacan via Shannon calls entropy. If it is somehow possible to introduce information into the circuit, information understood as that which makes a difference (Lacan offers the classic example of Maxwell's demon able to stop slower atoms from moving), the energy level rises because what would have been lost previously due to entropy (heat) has now been converted into work.[56] In the manifestos, one finds the wireless inserting itself in the circuit between man and machine, specifically the transmitting propeller, introducing information of a very specific kind, namely, that intuition or the creative spirit can take dictation over long distances when a certain speed is maintained between a writing hand and a recently transformed wireless imagination. An accelerated transmission between hand and imagination converts energy previously lost into *parole in libertà*, information for short. Information ought not to be confused with meaning, that is, the difference between what one could say with what one does say; instead, it measures "one's freedom of choice when one selects a message. The greater the entropy, the greater the freedom of choice and the greater the overall information that may be communicated."[57]

It is difficult to overestimate the importance of this piece of information (and secret command) for a history of wireless systems. *Parole in libertà*, the condensed, telegraphic output of the war correspondent that registers the Brownian movement of the hot zones, is the result of introducing information that limits (or extends, depending on one's point of view) what is experienced and written down to associations that can keep up with the infernal wireless command to incessantly exchange sense perceptions for their scripted form. A tentative history of wireless systems will view the manifestos as exhibits of entropy of the same system and then show how attempts to fasten a self-generating aesthetic strategy on to a semi-agent named Marinetti are conditioned primarily by a forgetting of technical media and the role they play in the emergence of free words, shaped writing, and onomatopoeia. Furthermore, a history of partially connected media discloses how the wireless command to speedy transmissions successfully distributes bodies according to the dictates of war, enabling an exchange of intuition that unites them in a closed circuit. The wireless increases its reach into previously unsounded realms, namely, the psyche and the senses, and its effect will be to send sense-registering machines (the invaded and extended body) to war. The invitation to poets to seek out the region's hot zones in search of untried *parole in libertà* was a first indication. An artillery barrage that requires a wireless transmitter if it is to be recorded will get one in the former poet in the same way that a propeller inclines Marinetti's head to its dictation. Tanks, machine guns, and airplanes, in a word, the machinery of war, act in concert with the wireless as they represent a more accelerated arena of Brownian movement than the sedentary one to be found in the city.[58] The symbolic relation between objects and *parole in libertà* is found both in the city and on the battlefield, but for the Futurist the difference between war and the city is entirely one of degree: it is simply easier to reproduce the discourse of noise if you go where the noise is, and war is fundamentally a noisy and speedy affair.[59] The vehemence with which the Futurists supported Italian intervention in 1915 and their success in registering its movements in "art," when not dying or losing a limb in the process, confirm that the wireless's command to speed enrolls bodies in the service of war.

Deficiencies and the Media Specialist

In a world in which more and more communications traffic is conducted wirelessly, the growth of the wireless media specialist will be spectacular. Although McLuhan was not the first to demonstrate how technology alters sense ratios and patterns of perception (excluding Marinetti, Roman Jakobson's 1919 study of Futurism perhaps deserves that honor), McLuhan was largely responsible for the important notion that media begin with physiological deficiencies. His categories of hot and cool were nothing less than an attempt to localize technological effects on sense media: radio "heats up" the ears of listeners while the blind seem particularly adept at constructing typewriters. There are difficulties with McLuhan's formulation of hot and cool media specialists, particularly his failure to account for the wireless as a realm separate from radio and telephony that exhausts flows of data. Still, he does direct us to physiological deficiencies and their important relation to technology.[60] In the following paragraphs, I want to sketch the most important features of physiologically deficient media specialist F. T. Marinetti.

We already chanced on one of Marinetti's most important deficiencies, alexia, the loss of the power to read, when we uncovered the dictated quality of *parole in libertà*. The operator of one's own wireless imagination, Futurist for short, registers the sights and sounds of modern life in a writing hand but does not read or edit the transmission for obvious reasons: with the abolition of syntax, the random arrangement of nouns and the end of punctuation, there is simply much less on the written page to correct and rearrange. Two consequences ensue. First, Marinetti's dissociated limbs produce scripted associations by utilizing a wireless imagination that demands that these same words remain if not unspoken, then certainly unread. Marinetti says as much when he cites the repeatedly raised objection that *parole in libertà* are not reader-friendly since they require special orators to be understood. Unfortunately, he abandons the manifesto's insight into wireless imagination and its dictated words, seeking comfort in the rationale that "any admired traditional poem, for that matter, requires a special speaker if it is to be understood."[61] By so doing, he ignores the significant differences that obtain between the two when media wars break out. What we know of Marconi wireless operations in the period in question also shows a wireless operator unable to read the words he registers at the moment of transmission. Because he transposes electromagnetic sound waves into their written, alphabetized equivalent, he is more relayer than reader: there is simply no time for reading given the command to speed that the wireless imposes.

The second consequence is more broad-ranging as it concerns the use of onomatopoeia or, better, Marinetti's failure to spell correctly. In the manifesto "Destruction of Syntax," he underscores the new spelling rules for *parole in libertà*, by demanding, in italics, a *new orthography* and a *free expressive*:

> Thus we will have the *new orthography* that I call *free expressive*. This instinctive deformation of words corresponds to our natural tendency towards onomatopoeia. It matters little if the deformed word becomes ambiguous. It will marry itself to the onomatopoetic harmonies, or the noise-summaries, and will permit us soon to reach the *onomatopoetic psychic* harmony, the sonorous but abstract expression of an emotion or a pure thought.[62]

A writing minus orthography is one that knows no best usage; only a pure writing creates the conditions necessary for psychic onomatopoeia, defined as the expression of pure thought and emotion. Spelling will go the way of syntax, conjugated verbs, and the adverb, mere clutter on the highway to a more abstract symbolic encoding of the real. Interestingly, previously correct spellings of words also undergo a change, a deformation that occurs in the act of transmission by a wireless imagination to the writing hand. They are the material proof that a transmission is taking or has taken place. The wireless inscribes them with the signs of transmission; naturally, a hand that writes independently of any ego writes with deformed, misspelled words.

A further deficiency joined to the concrete expression of pure thought awaits the nonreading wireless operator. Compare the following:

Just as aerial speed has multiplied our knowledge of the world, the perception of analogy becomes ever more natural for man. One must suppress the *like,* the *as,* the *so,* the *similar to.* Still better, one should deliberately confound the object with the image that it evokes, foreshortening the image to a single essential word.[63]

As a child I left out words such as "is," "the," and "it," because they had no meaning by themselves. Similarly, words like "of" and "an" made no sense. Eventually I learned how to use them properly, because my parents always spoke correct English and I mimicked their speech patterns. To this day certain verb conjugations, such as "to be," are absolutely meaningless to me.[64]

The first passage is Marinetti detailing how corresponding analogies will suppress the middle terms between the object and the image it evokes in the condensed word. The second is the account of a life with autism by Temple Grandin, world-renowned designer of livestock-handling facilities. The two process nonvisual information in remarkably similar ways (Grandin's description of her condition appears under the heading "Processing Nonvisual Information"). The question for both is how to turn objects into meaning. The first step is to remove word clutter: the autistic child leaves out articles, prepositions, and pronouns because they do not offer a direct path to meaning. Marinetti's wireless imagination operates along the same lines, removing words that hinder the easy transmission of sense data into their written analogue. For him, the joining of words with their doubles no longer requires the linking verb, so that one will eventually read "man-torpedo-boat" and "women-gulf," etc.[65] Nouns present little difficulty for the autistic child and Marinetti because they are more closely related to pictures, which in turn lend themselves to highly associational thought patterns ("the easiest words for an autistic child to learn are nouns," she offers at one point). Indeed, the more severely impaired the autistic, the more highly associational the process: "Lower functioning children often learn better by association, with the aid of word labels attached to objects in their environment. Some very impaired autistic children learn more easily if words are spelled out with plastic letters that they can feel.[66] Marinetti's doubled nouns arrive at the same end by attempting to *fondere* the object with its evoked image. *Fondere* is itself evocative as it both describes the process of liquefying a solid and uniting together two or more things through amalgamation.[67] Amalgamation for the intrepid Futurist occurs when a wireless imagination increases the grades of distance that separate analogies:

They have compared, for example, a fox terrier to a very small thoroughbred. Others, more advanced, might compare that same trembling fox terrier to a little Morse Code Machine. I, on the other hand, compare it to gurgling water. In this there is *an ever-vaster gradation of analogies,* there are ever-deeper and more solid affinities, however remote.[68]

The deep and solid relationship analogy maintains with its adjacent double is the result of liquefying the boundaries between the object and the image that the condensed *parole in libertà* evoke. Appropriate and inappropriate associations are beside the point for an aesthetic strategy that condenses image and object in the essential word. Grandin confirms the same for the autistic child: "When I was six, I learned to say 'prosecution.' I had absolutely no idea what it meant, but it sounded nice when I said it, so I used it as an exclamation every time my kite hit the ground. I must have baffled more than a few people who heard me exclaim 'Prosecution!' to my downward-spiraling kite."[69]

Grandin's "Kite-Prosecution!" and Marinetti's "fox terrier–gurgling water" both attest to the nonsense that advances beyond meaning to the body that produces it. This is the case as well for blows to the head and certain kinds of neurological damage that lead to aphasia or inappropriate word associations. Indeed, Grandin quotes recent studies of "brain-damaged patients" to demonstrate how injury to the left posterior hemisphere can stop the generation of visual images without impairing verbal memory.[70] Marinetti's simulated wireless imagination operates similarly, creating the conditions for physiological deficiencies that can short-circuit the part of the brain that works on verbal thought. If registered and transmitted quickly to a writing hand, the noise of modern life can keep Marinetti busy indefinitely, which is precisely the problem for people with more severe autism—stopping the endless associations that seemingly run on automatic in their minds. Fortunately, Grandin is able to "get her mind back on track," but when associating, her mind becomes a video that jumps from image to image or a CD-ROM disc that randomly shuffles its program. Marinetti's wireless imagination also cracks the whip of association with what he calls at one point the "inexhaustible richness of images" produced by his wireless imagination in conjunction with the various data flows it registers.[71] Physiological deficiencies mirror the effects of simulated wireless practices.

Interesting enough as a case study of the subterranean connections between physiology and technology, the examples of autistic and Futurist information processing interest me as well for what they tell us about how the wireless alters notions of poetry and imagination. If simulated wireless operations uncouple writing from a controlling consciousness, so as to allow its practitioner to enter into synchronous vibration with a wider spectrum of visual and acoustical frequencies, then the wireless command to velocity has successfully inserted itself into areas previously considered off-limits. The wireless does not just alter sense ratios or patterns of perception à la McLuhan, challenging and extending the responses of our senses and ourselves to become what we hear or behold. Para-phrasing the opening of "Destruction of Syntax," the wireless models under-standings of poetic inspiration. The manifestos demonstrate that the effects of inspiration and the creative spirit working blindly on the poet can be simulated via wireless practices that function as physiological or neurological disorders. Whether a poet installs a receiver in his brain or begins to shuffle randomly sense registrations autistically, the result is the same: imagination is revealed as another information processor and poetry as another medium among several available at the moment of transmission. This was already apparent in Sergi's experiments when hearing the sounds meant manufacturing a space between them. A wireless imagination also registers the space between sounds, but it is not content with simply one stream of information. It expands its data-flow domain and does Sergi's machine one better: it symbolically encodes them uti-lizing a writing, alphabetized hand. At this point, it is tempting to circle the wagons, searching for traces of an imagination that has not been uncoupled by a wireless technology bent on its end. Marinetti is no different, hanging on as he does to the exposition of thought as opposed to, say, the nonsense of Doctor Schreber. But once the command for speedy transmissions is sent, little remains to be done except to account for the absence of the usual literary suspects: the lyrical I and inspiration among them.

Fascist Transmissions

What does the insight into wireless commands and sense fields tell us about fascism and technology? Or to be more precise, what does the intersection of the wireless and aesthetics in the manifestos indicate about their relation to fascism? The present inquiry suggests two approaches. The first attempts to destabilize what we understand by "user" of technology by detailing wireless practices that standardize response thresholds to nonmeaning—the user is measured and, recalling Lacan, transformed into just another piece of information. As the wireless expands its bandwidth to include the higher frequencies of the vi-sual, the fast-disappearing user is found in the distance between analogies and the speed with which his or her wireless imagination dictates condensed *parole in libertà* to an unconscious hand. Greater or lesser fascination with film, the gramophone, the wireless—in a word, media—miss the mark as they continue to presuppose areas that do not fall to a technological standardization of sym-bolic encoding. By undermining the notion of user, the approach shows how indebted the fascinated fascist argument is to the oppositional logic of autono-mous subject versus thing that is often found agitating in definitions of the fas-cist as an armed and mystified user of technology. It may well be that the fascist along with his *marconista* fellow traveler may only be exhibits in a history of wireless media systems. Thus, the present chapter demonstrates how important imaginations and hands are in the ever expanding relation of wireless and operator; aerial viewing and theophanic vision are the bonuses of a more efficient symbolic encoding of the real.

The second approach returns our attention to the aestheticization of forms outlined in the previous chapter. Aestheticization in the manifestos occurs early on in the *parole in libertà,* which become the secret code of an initiation that addresses potential Futurists on how to install wireless imaginations. In an aside in "Destruction of Syntax," Marinetti describes a sort of telepathy between the public and poet; words may be coded so as to short-circuit the traditional pathways of understanding: "They can make themselves understood with half a word, half a gesture, a glance. So the poet's imagination must weave together distant things *with no connecting strings,* by means of essential *free* words."[72] The wireless command to immediate sensory registration finds its corollary in the immediate transmission between poet and public expressed in the half word or, better, the wireless code of the day. The speedy transmission will later link up with the successful amplification of the speaker's voice, creating a chain of technologies, the early determinations of which interconnect in the 1912 "Technical Manifesto" and in later formulations of fascist modernism.[73]

That it is a small step from poet to *il duce* was already apparent in D'Annunzio's Fiume orations. The same is true in Marinetti's works where the wireless creates a space in which clear-cut distinctions between art and politics no longer abide. There it becomes an object of art and an emblem of a new mode of registering sense data. The proof is everywhere. The command to rapid transmission in the manifesto modifies patriotism to become the "heroic idealization of the commercial, industrial, and artistic solidarity of a people."[74] War is altered by the wireless, making it faster and more compact, accelerating it to such an extent that it can only be measured as a vector of energy residing in a people: "A modification in the idea of war, which has become the necessary and bloody test of a people's force."[75] The wireless, like every other technology before and after, goes to war. What the wireless makes possible on the battlefield, the rapid transmission of orders, it successfully extends in the command to patriotism and attunement between the collective and *il duce.*

Yet the coded and condensed half word that shrinks the distance between poetry and politics as it opens ears to the oracle always comes with a price: no blood offering, no oracle.[76] The blood flows freely in the manifestos as so many commentators are quick to point out, but the sacrifice is rarely seen as part of an exchange for effective contact with the oracle. And the blood that has to be contributed in real life cannot be substituted: once given, there is the real risk that the wireless, like so many revived ghosts before it, will begin to assert a will of its own.[77] Both approaches to fascism and the wireless therefore arrive at the same conclusion, albeit through different means: wireless media collaborate with fascism. Recall how D'Annunzio's speech at Fiume looked forward and backward to a present when poetry and the wireless transmission would result in the pure unmediated act. Marinetti's encounter with the wireless moves us more securely in the direction of D'Annunzio while forcing us to give the wireless ever more its due. As the command to velocity issued by a wireless imagination shifts bodies at previously "unheard of" velocities, the wireless transmission will begin more and more to coincide with war itself, meeting up, if Thomas Pynchon is to be believed, in the wirelessly guided V-2 rocket that explodes in a movie house.[78] Marinetti's encounter with technology demonstrates that wireless media and fascism work together by submitting bodies to greater speed.

The Wireless Voice Transmission

As the present chapter closes, a number of questions persist. Is the wireless transmission simply a nonvocalized, written system of notation? What happens when the wireless no longer requires an unconscious hand for the registration of its transmission? Does the wirelessly transmitted voice do justice to its Fiume and Libyan pedigree? To answer these questions, we need to jump ahead some thirty years, to a time when voice transmissions were quite literally the order of the day. Ideally, our choice of study should be one operating in two media, preferably a writing and wireless specialist since a comparison of media and incompatible data channels would make clearer the often complex relationship that holds whenever partially connected media come into play. Moving to center stage, Ezra Pound's *Cantos* and his "treasonous" broadcasts over Radio Rome during World War II offer further evidence of a wireless command to velocity and a fascist aesthetization of immediacy that submits bodies to war. The following chapter shows how the wireless succeeds in generating not only words and analogies but voices as well. The outlines of a media situation altered by the reality of collective long-distance voice transmissions helps specify the debt Pound's "Radio *Cantos*" owe wireless technology in the age of Marconi.

4

Pound's "Radio Cantos": Poetry and Data Retrieval in the Age of Marconi

THE LINE that runs between the source of Marinetti's *parole in libertà*, the wireless imagination that exhausts associations in script and the receiver, the Futurist able to hallucinate a meaning in the condensed transmissions, is for all its speed rather rudimentary. First, the writing hand is not the most efficient channel for carrying the signals the wireless imagination picks up: a direct line to the mouth is needed since the velocity of transmissions to a speaking voice is significantly faster than those to a writing hand. Second, early wireless transmissions registered in writing have difficulty switching frequencies, perhaps because handwriting flows: when war correspondent Marinetti records the battle of Adrianopolis, for instance, *parole in libertà* are limited to the registration of nonhuman sounds. Voices are absent because the fine-tuning necessary to pick them up and register them is lacking. Indeed, the Futurist aesthetic remains circumscribed in its encounter with the wireless: as late as 1933, with the publication of the manifesto "La radia," the possibility of voice transmission is ignored in favor of *la radia*'s general capacity to capture, amplify, and transfigure Brownian movements. No distinction is made between fine-tuning animate and inanimate wireless transmissions.[1]

Two technical developments—the progressive strengthening of transmitters and the fine-tuning of near frequencies—spell the birth of wireless voice transmissions while providing a means for identifying how poetry, represented here by one of Pound's cantos, behaves under altered media conditions. Therefore, I dedicate the first section of this chapter to an examination of the wireless in World War I, the battle for cables that preceded the outbreak of hostilities, and its emergence as a primary transmission medium. On the one hand, the wireless improved the operations of war by its very speed; on the other hand, it required that certain measures be taken to prevent enemy interception. One such measure was cryptography, the design of ciphers and codes to make a message unintelligible. This development, linked as it is to writing, further specifies the nature of the wireless dispatch in its lack of signature, the difficulty of localizing the source and destinator, and the transmission's heightened condensation. The wireless no longer needs the experiments of psychophysicists to furnish the spacing needed to create listening subjects, or to produce words freed from addresses.

This chapter's middle section describes what happens when the wireless enlists voice as a channel for its transmissions. By comparing two guides for successful wireless transmission, Everett Gordon's instructions to the U.S. Signal Corps from 1920 on the proper rhythm necessary for a wireless spark signal to be transmitted and Rudolf Arnheim's coachings from 1936 on how to announce properly over the radio, I show how both install blindness as a fundamental condition for pattern recognition on the wireless. I argue that the symbolic spacing enacted in both wireless transmissions, be they of the voice or spark variety, is the medium's most significant feature. This is not to say that the acceleration the wireless voice transmission represents was of negligible impact, only that, in the move from wireless telegraphy to wireless radio, what remains essentially wireless is maintaining a certain spacing, what the *marconisti* of the time refer to as possessing "good fists."

Once I have joined spark to voice transmissions, I turn to two developments that extended the reach of the wireless: the fine-tuning of signals and the emergence of media linkups. Fine-tuning grew out of American and British attempts at the end of the war to pick up high-frequency German transmissions. The device in question, the superheterodyne, made audible formerly inaudible high-frequency transmissions in a process whereby a first signal was mixed with a second to produce an audible third. In practice, this meant that a listener could fine-tune any number of stations across a large bandwidth, even those close in frequency: where the ear could not distinguish between near transmissions, the superheterodyne could. It was no longer simply a question of receiving the signal but of tuning into stations: inaudible or weak signals could be fine-tuned. The second wireless extension concerns the growth of what Friedrich Kittler and John Johnston term systems of partially connected media. By the end of World War I, the wireless was distributing words and sounds, emitting and connecting them across ever greater distances. To be sure, transmission quality was poor, but that was largely caused by the transmission limits of the relay in question—novice hands could code, on average, only eight words per minute. Despite the slow connection speed, the transmitted form of the word successfully set out its potential for other couplings, in this instance with voice; the result was an increase in transmission quality given the relative detail of speech when compared to tapped wireless code. What emerges from my reading is something akin to the reproduction of a relay among voice and storage media. The wireless voice hookup confirms the previous network that connected emitting and receiving stations while simultaneously generating an extension of the same system's operations.

Only after entertaining the possibility of a wireless system do I examine how the wireless accelerated and condensed poetry, forcing it to develop a capacity for registering transmissions on more than one frequency. My choice of Ezra Pound as an object of study may seem all too obvious considering Pound's history with radio and the truly staggering amount of material devoted to the relationship.[2] But his appearance in a study of wireless media is dictated by three further considerations. First, because Pound worked in both poetic and wireless media, a comparison of his production in both may shed light on the precincts in which the wireless and poetry overlap. Second, few poets were as attuned as Pound to media, and I am not just referring to the radio. Pound frequently went to the cinema; he was an early devotee of Tarzan, Mickey Mouse, and Fred Astaire.[3] Yet, he was not just a consumer, having planned as early as 1931 to produce a documentary film on fascism. Equally significant, Pound was one of the first to compose both poems and letters almost exclusively on the typewriter: for instance, most of Pound's letters to Joyce are typescripts. The command to "write it down" that I first noted in Marconi as one of the most significant characteristics of the wireless transmission returns in Pound, and as befits the growing complexity of partially connected interwar media systems is issued not to a writing hand but to the symbolic spacing a Corona typewriter makes possible (and we might add demands). Pound becomes just another piece of information in the relation between the source of his poetry, the voices he picks up, and the typewriter that mechanically transposes voices to paper. That Pound also availed himself of his daughter, lover, and wife as typists and stenographers for the *Cantos* does not go unremarked since it points to the position women occupy as possibly the ultimate addressee of Pound's production. When I examine Pound's transmissions on Radio Rome in chapter 5, the media system expands to include voice registration on source disks that allowed for rebroadcast anytime. In this chapter, however, I will be concerned with how the typewriter connects the wireless with an ever-widening network. In my reading, Pound becomes part of an information network, whether he is transmitting in writing or over the wireless: his poetry, his radio broadcasts, and his pedagogy presuppose the wireless media of the period.

Pound's injunction to model one's reading of the *Cantos* around the development of radio informs the final discussion. There is the oft-quoted letter to Ronald Duncan of March 31, 1940, in which Pound claims to have "anticipated the damn thing in the first third of the Cantos," but two other examples are more to the point in a chapter dedicated to the specificity of the wireless.[4] In a letter to his father from 1924, Pound compared his method in Cantos 18 and 19 to the radio this way: "As to Cantos 18–19, there ain't no key. Simplest parallel I can give is radio where you can tell whose *[sic]* talking by the noise they make."[5] The second comes from an article Pound wrote for the Genoa-based journal *L'indice* in October 1930. Describing how translation and poetry ought to mirror developments in media, he writes: "But one can try to seek out this brevity, this style, not really telegraphic, though it does belong to the epoch of the telegraph. In the end, clarity and vigor of thought will make the new style."[6] The "noise they make" and a style that belongs "to the epoch of the telegraph" provide the coordinates for my reading of Pound in the age of Marconi, marking where poetry registered technical advances in voice transmission and increased speed and condensation. I contend that *A Draft of XXX Cantos* occurs under the sign of the telegraph and the radio, longhand for the wireless.

Finally, I ask how an examination of Pound's encounter with the wireless aids our understanding of the nature of the encounter between user and machine. The body of the poet returns as a locus for wireless effects, and here I lean on a number of Pound's psychologists for their diagnoses since they, more than most of Pound's critics, recognize the intimate relation between associative method, wireless transmission, and physiological deficiency. To speak of deficiencies and poetic production returns our attention to fascism and to the important questions anticipated in previous chapters: the fascist as a subgenus of the wireless operator who bends his ear to the oracle while picking up distress signals, fascism as an information loop that successfully extends wireless capacity by connecting lower echelons to a command central, and the poet as a cryptographer commanded to short-circuit the traditional pathways of understanding between *il duce* and a set of initiates tuning into the transmissions. Pound's wireless writing demands that we read him as a cryptanalyst would, overhearing his transmissions and deciphering his codes, for that is what a wireless transmission requires: we are noting the noise, comparing messages, and finding a semicoded transmission that, with some work, may provide a key to reading Pound.

Media Wars and the Wireless

Wireless Addresses

To locate poetry among a number of competing transmission media, a survey of the media landscape before, during, and after World War I will be necessary. For the most part, cables and wireless media were engaged in a series of skirmishes in the early part of this century, each trying to "successfully clobber the other" in Marshall McLuhan's words. The last third of Luigi Solari's biography of Marconi reads like a geography of media flashpoints: the Balkans, where the wireless promised to break the network of German-owned cables; Germany itself, whose scant 7.5 percent of the world's cables was clearly insufficient to guarantee that its messages would reach their destination; and Great Britain, which despite (or on account of) its position as the world's leading cable center became the site of so many battles among competing media. Solari spreads Marconi's wireless gospel far and wide, aided by an audience that knew just how dependent it was on someone else's cables for communication. On the one hand, the wireless promised to end that dependence by eliminating the need for physical linkage by wire, while it sped up "communication between headquarters, joined through the ether units that could not connect by wire because of distance, terrain, hostile forces or rapid movement . . . and eased the economic burden of producing immense quantities of wire." On the other hand, "the public, omnidirectional nature of radio transmissions, which makes wireless communication so easy to establish, also allows for easy interception. It was no longer necessary to gain physical access to a telegraph line behind the enemy's front to eavesdrop upon his communications."[7] The very nature of the wireless transmission, the lack of wires that made it so appealing to noncabled powers, was also its chief drawback. If the wireless was to compete successfully against cables, a solution to interception had to be found. Not to do so was clearly a matter of life and death as the Italians sadly discovered at the Caporetto debacle of 1917.[8]

The competition among transmission media was staged on two fronts: in the wireless capacity to keep up with armies on the move and in the vulnerability of cables to cutting. In the first, the speed with which men and equipment moved, at least in the initial stages of the war, required a medium capable of connecting two points that were continuously shifting positions.[9] The opening German campaign in France is instructive. Acutely aware of the Japanese success with the wireless in Manchuria in 1905, the German army had decided to go into the field with only wireless and telephone, making no provisions for alternate means of communication. Once they passed beyond the wires of their telegraph network, they were forced to begin transmitting wirelessly. These signals, intended for German ears, were intercepted by the French, demonstrating again that intention ought not to be confused with the structure of a medium. The French deciphered the messages easily enough, as they had been intercepting German communications for several years before the war; by 1914 they were the most accomplished cryptologists in the world. No little help was provided by German wireless operators, who inexplicably maintained the same wireless key for eight days over the entire Western Front; it was only a matter of time before the wireless transmissions were deciphered and the next German move prepared for.[10] As the war quickly moved to attrition by the spring of 1915, the number of German military wireless messages decreased dramatically, only to pick up again with the German offensives of 1917. Over the course of the war, the French intercepted more than one hundred million words, "enough to make a library of a thousand average-sized novels."[11]

The ease of wireless interception requires some provisional commentary. First, note the deceptively simple fact that the wireless has no need of wires to transmit to another single place in space. Since the wireless signal spreads out in all directions, its messages lack a specific address. The lack of address has often been conjoined to a universal one: it is a transmission that addresses everyone and no one in particular, which may account for its entanglement with spiritualism over the years.[12] In other words, the wireless is not a telephone, that is, it is not directed to a particular addressee; its messages are transmitted ubiquitously, increasing the chances of interception and putting a premium on the development of cryptological techniques.[13] Any number of encodings and ciphers are possible, but they are merely attempts to supplement the lack of an address, to mark it as a transmission to someone rather than to all and no one; they counteract the lack of specificity inherent in every wireless transmission.[14] One recalls D'Annunzio at Fiume sending a marconigram that would make the world's antennae vibrate, hiding destinations and addresses under the cover of universality. Cabled transmissions, either via telephone or telegraph, require less encoding or ciphering since wiretapping can produce intercepts only at irregular intervals. This is because cables, at least at the turn of the century, were always station-to-station.

Second, the lack of an address entails a difficulty in localizing the transmission's provenance. Call signals will later be developed, and with it the broadcast band, to remove doubt about who or what is transmitting, but their function differs little from encryption techniques to the degree that they too supplement a lack, here of a localizable addressor. They ease the burden of uncovering the source of the transmission, "promising a unification of signal with source."[15] Yet, the promise is illusory. It is precisely because the place of emission is not fixed that one hears a signal or voice over the wireless. Just as soon as one gets a fix on the place of emission, strictly speaking, it will no longer qualify as a wireless transmission.[16] Goebbels would make much of this wireless capacity for nonlocalized transmissions. After surreptitiously recording announcements made by French and Belgian speakers over the radio in early 1940, he used the same voices to mask the identity of the Nazi secret transmitter, transmitting, naturally, on the same wavelength.[17]

High-Frequency Alternators

Let us return to this century's early media wars. In the period immediately preceding 1914, cable and noncable powers began projecting wireless networks that could circumvent cables. The Germans constructed their wireless network at Nauen near Berlin, while the French answered with a system to combat the Germans, especially in transmissions to South America; Italy's network essentially connected stations along the Mediterranean rim. The United States remained blissfully ignorant of its cable dependence until the outbreak of war when its need for new channels of communication became desperate. At the forefront of wireless networks, however, was Great Britain, which in 1913 had projected a worldwide wireless project of her own, the "All-Red Chain." Yet the wireless remained a paper tiger for several years: the distance between stations was enormous; Marconi's spark transmitter was a decidedly undependable instrument with which to reach agents in the field; and it was still notoriously difficult to pick up stations broadcasting on near wavelengths. The war, however, would dramatically alter the media landscape.

If the wireless cuts wires, only to rejoin them at ever greater distances, then clearly we should be on the lookout for lines severed during the war. Indeed, one of the war's first media casualties was the German transatlantic cables to New York, cut in early 1915 at the Azores and Liberia. The result was British control of "virtually all communication between America and the Central Powers." Because cable messages had to pass through England, "instead of censorship at the source there was censorship in transit."[18] Into the breach created by cut cables stepped the first wireless extension of the war, the high-frequency alternator. Or rather two alternators, the Alexanderson high-frequency alternator in the United States and the Goldschmidt high-frequency alternator in Germany. These generators, the size of locomotives, were hooked into the output circuit in order to transmit Morse code and had the added attraction of not catching fire like so many high-power spark transmitters before them. In the United States, the British Marconi Company installed the Alexanderson alternator, named for General Electric engineer E. F. W. Alexanderson, its inventor, at its New Brunswick station in early 1917, shortly before the station was taken over by the U.S. Navy.[19] Soon the station was communicating directly with Paris and Rome. In May 1918 an even larger alternator was installed at New Brunswick, which transmitted President Wilson's Fourteen Points directly to Germany's high-frequency alternator at Nauen. It also maintained ship-to-shore contact with the presidential ship, the *George Washington,* during President Wilson's trip to Europe following the armistice.

The significance of the alternator for wireless transmission lay singularly in the increased strength of the transmitted signal. By augmenting its range, picking out atmospheric S's from wirelessly transmitted ones was now easier given the boost of power the alternator provided. With a single high-power generator and a transcontinental chain of relay stations, command could theoretically issue orders to forces abroad independently of any cable technology. This development ought not to be underestimated. As Stephen Kern suggests, "new weapons made it possible to increase the space between enemy soldiers, while electronic communication could extend the distance between sectors of coordinated operations and between commanding officers and their men. In several respects World War I was conducted at unprecedented long range."[20] Whereas Napoleon could not do without his seconds-in-command to relay orders, high-frequency alternators and the stronger signals they sent now meant that com-

An early example of a high-power spark transmitter at Clifden. Reprinted by permission.

manders could direct men to their deaths from a bunker, a lab, or a studio.[21] What's more, the large numbers of troops that had been required to control and persuade others to die could now be replaced by wireless technologies. The substitution of a rhetoric of presence with a wireless one, accompanied by the connotations of wireless magic I noted in chapter 1, did not happen immediately, of course. The Spanish Civil War and the political advisors that maintained each side's ideological purity, along with the presence of commissars for Soviet soldiers and National Socialist political officers for German ones in the next European war, remind us that the process was a slow one. Still, the days when one half of an army motivated and controlled the other half had passed.[22]

From there, it was a short step to what German armaments minister Albert Speer at Nuremberg identified as the key multiplier of Nazi power:

> By means of such instruments of technology as the radio and public-address systems, eighty million persons could be made subject to the will of one individual. Telephone, teletype, and radio made it possible to transmit the commands of the highest levels directly to the lowest organs where because of their high authority they were executed uncritically. Thus many offices and squads received their evil commands in this direct manner. . . . Dictatorships of the past needed assistants of high quality in the lower ranks of the leadership also—men who could think and act independently. The authoritarian system in the age of technology can do without such men.[23]

I will return to Speer's thesis when outlining a system of wireless interconnectability, but for now we are in position to consider the significance of the high-frequency alternator in a history of wireless extension. Assisted by the Alexanderson and Goldschmidt generators, the wireless extended the distance of its transmissions while simultaneously joining stations across them. By so doing, it extended war into new theaters: the British attack on Gallipoli in 1915 was unthinkable without the wireless, as were the German Zeppelin bombing raids over England begun in August 1914, which reached their apex in 1917 with the development of two-way, air-to-ground wireless equipment. And just as the Futurists became war correspondents covering generated theaters of war in Bulgaria and Libya, now millions of men were given the same opportunity across Europe.

The Wireless Voice Transmission

The increased strength of the wireless signal had another important effect. Alexanderson and Goldschmitt alternators could conduct voice transmissions more easily and with less risk of explosion than earlier versions of the wireless. To be sure, voice transmission had been attempted as early as 1906, but the power requirements of the early transmitters made carrying voices impractical. The alternator and soon after de Forest's regenerative circuit and audion made transmission easier, though if one historian of the wireless is to be believed, the extension was already contained in the period's extant cable technologies. "The truth of it was that it wasn't a difficult step inasmuch as all of the electromagnetic and mechanical elements required were, for the most part, already being used in telephonic and telegraphic applications. It was only a question of moving from one order of phenomena to another, which principally required further refinements of the transmitting and receiving stations being used."[24] Other technical accounts from the period confirm that often little or no distinction was made between signal-making instruments, be they the telephone, a galvanometer, or a relay of some kind.

This raises an important question. What is the significance of the lack of cables for the propagation of voice, that is, how does hearing and transmitting a voice on the wireless alter the parameters of the early wireless spark transmitter? The question is problematic, for differences inhere between spark and voice transmissions, most significantly the greater amount of information contained in the voice, where information is understood à la Weaver as the measure of one's freedom when selecting a message. Clearly, the dimension of greater selectivity in voice is a concern, but it is not the only one. What if the question concerns, rather, the status of tapped wireless transmissions and whether they function as voice transmissions do? If they do, then we are well on our way to identifying voice and writing as "orders of phenomena" of the wireless. This is an essential preliminary to meeting Derrida's injunction that there is no speech that is not already writing and hence that does not fall to *différance*.

To understand what I mean, I want to gloss an early text of Lacan's and assume an empirical conception of speech and writing. In "The Creative Function of Speech" from 1953, Lacan recounts the story of Odysseus's companions and their metamorphosis into swine. He asks how one might recognize whether "a grunt which comes to us out of the silky mass filling the enclosed space of the pigsty is speech." After rejecting the possibility that emotion or feeling expressed in the grunt constitutes speech (it does not, for it lacks complexity), he argues that the grunt becomes speech when the swine want it to be believed as theirs. In Lacan's words, "The pig's grunt only becomes speech when someone raises the question as to what it is that they want to make you believe. Speech is precisely only speech inasmuch as someone believes in it. And what through grunting, do these companions of Odysseus turned into swine want to have us believe? —that they are still somewhat human. . . . Speech is essentially the means of gaining recognition." If that is so, then the swine may be said to grunt the following: *"We miss Odysseus, we miss his being with us, we miss his teach-*

ing, what he was for us in life."[25] Lacan goes on to distinguish between a communication that transmits, in his words, a mere "mechanical movement" and the dimensions opened up by recognition: "But, as soon as it [the pig] wants to have something believed and demands recognition, speech exists."[26]

Admittedly, Lacan's analysis of the function of speech is intended as a link between the metaphorical base of every kind of semantic usage and the structure of the analytical space; grafting it on to a conception of wireless voice and writing risks occluding these considerations. Still, his distinction is helpful in circumscribing the dimension of speech in wireless spark and voice transmissions. To do so, we need to identify the status of the elusive "something believed." What do the swine ask the listener of their grunts to "believe"? Lacan translates the grunts for us: they miss Odysseus and what he represented for them in life. Note that it is not wanting to possess a past that ushers in the realm of speech but the wish to have "something believed" and the demand for recognition that moves the event into speech. Neither is misrecognizing the provenance of the grunt as a condition for whether speech may have occurred. In other words, recognizing the swine's grunts as speech does not entail making them over into the human. The swine that wants its "something believed" does not refer to a localized entity speaking from within the swine but to a demand that its act of speaking be recognized as such.

Translated into terms congenial to the wireless, spark and voice transmissions occur in the dimension of speech to the degree their source seeks recognition. When the provenance of a transmission is recognized, whether it be the commanding officer on the Western Front or *il duce* at Fiume, speech may be said to occur. When the provenance of the transmission is not recognized, we move in a different world of dots and dashes and ghostly voices whose nonlocalized presence makes speech impossible. These are extremely important considerations for reading Pound's wireless writings and radio speeches; here I want to focus on their significance for wireless transmissions. Following Lacan, a want for recognition constitutes speech on the wireless. Although he does not list them for the swine, one can imagine what they might be for the wireless: outright identification in call letters; a stronger signal; repeating transmissions; pitching voice so as to make it more recognizable to its listeners; and, in spark transmissions, coding hands that maintain a certain recognizable rhythm.[27] Bear in mind, however, that the constitution of speech in recognition is distinct from the question of spacing I outlined in chapter 3. Spacing (primarily between dots and dashes but not limited to them) is a condition for generating meaning and creating listening subjects willing to follow orders. The "demand" for recognition occurs on the source side of the transmission.[28]

I want to illustrate the above with two examples. The first comes from a typewritten manual from 1920 on how to send properly over the wireless. Here, Everett P. Gordon obsessively instructs his wireless operators on maintaining a deliberate rhythm in their transmissions, as the following excerpt indicates:

> Above all do not attempt high speed transmission. Do not send faster than eight words a minute until you are able to copy at that rate. While transmitting always keep in mind the man who is receiving you and endeavor to give him as little trouble as possible. Nothing is quite so disconcerting when receiving as poor spacing by the transmitting operator. Poorly executed characters may be read by a skillful operator, but when in addition, they are poorly spaced, the task of receiving becomes well-nigh impossible.[29]

To aid those receiving, "at the beginning of every transmission an 'ATTENTION CALL' is sent. A 'PERIOD' denotes the end of a sentence, the 'CROSS' that the message is completed. A special signal is used to denote the 'END OF ALL CORRESPONDENCE.'"[30] Once wireless operators gained proficiency in spacing, they joined the ranks of those possessing "good 'fists'":

> Proficiency with a telegraph key goes far beyond simple knowledge of the code to the speed and rhythm of the sender. Those accomplished with the key were said to possess good "fists." Indeed, in the early days of telegraphy and wireless, practiced listeners knew the sender by the particular rhythm of the key taps, what operators termed "swing," just as those with well-trained ears may tell the difference between a Horowitz and a Rubinstein at the piano.[31]

A certain recognizable rhythm, the "good 'fists'" of those accomplished at the key, represents a wish to be recognized as the provenance of the transmission. No meaning that may be attributed to the decoded transmission makes it speech; rather, it is the recognition that is factored into the swing or rhythm.[32] Repeating the rhythm maintains recognition.

Consider next these coachings from Rudolf Arnheim in 1936 on how to send properly over wireless radio:

> A step in the right direction is to dehumanize the announcer as much as possible. Nothing should be heard of his bodily existence in the studio, not even the sound of his footsteps. Even his voice, the only thing that is left of him in the damped room where he has to be painfully quiet, must have no character, nothing peculiar or personal, but must only be distinct, clear and pleasant. For the function of the normal type of announcer today in no way differs from that of print, which should be inconspicuous, agreeable, easy to read and nothing more.[33]

Arnheim's artful instructions on how to dehumanize the budding announcer are worthy of further comment, which the reader will find in the following chapter. For now, I want to focus on the scene of transmission with its pain-fully quiet announcer who, when speaking clearly and distinctly, slips into the twilight zone of absence and presence, "nothing more" than black letters on a white background. What is the difference between Arnheim's insistence on a clear and distinct mode in voice transmissions and the properly spaced dots and dashes transmitted by trained, "swinging" fists? Clearly, Arnheim is de-scribing a different enterprise than Gordon's, in this case how to announce over the state-run radio of 1930s Germany; wireless voice transmissions were already the order of the day. Yet a comparison between the two is most instructive for pinpointing where they overlap. First, both transmissions exclude sequences of noise by utilizing a proper symbolic spacing between meaning sets. Call the spacing "creating no trouble," "style," or "pleasant character," it represents only the switchboard of scannings and selections made among real data flows.[34] A further point is easily missed. Spacing serves not only as a condition for gen-erating meaning but also for recognizing the source since both transmissions demand that their sources be recognized through the "swing" with which their operators work. The acute significance of spacing that I first noted in Sergi and then tracked in Marinetti's *parole in libertà* appears to be characteristic of the wireless voice transmission.

If, paraphrasing Lacan, the greatest light is also a source of obscurity, then the installation of blindness may help determine the source of a wireless trans-mission. In the two selections, the listener is asked to hear with his ears and not read: both Gordon and Arnheim insist that the wireless operator and an-nouncer must forget their eyesight. Gordon, for example, specifically forbids his students from learning the characters visually: "The sound method . . . trains the student in such a manner that upon hearing a character he translates it immediately and instinctively into the latter for which it stands. He does this subconsciously or instinctively . . . and does not find it necessary to visualize the dots and dashes."[35] The wireless operator is one who confines his thoughts "to the sound of the signal," which is not surprising given what we know about the relationship between *marconisti* and the wireless signal. Arnheim offers the same advice under the subtitle "In Praise of Blindness":

> One can dispute whether the aural world alone is rich enough to give us lively
> representations of our life, but . . . no further doubt is possible that the visual
> in any case must be left out and must not be smuggled in by the listener's
> power of visual imagination. Statues must not be subsequently given a coat-
> ing of flesh-tints, and a wireless broadcast must not be envisaged.[36]

The symbolic spacing enacted in wireless transmissions allows for meaning to be hallucinated—there have to be spaces between dots and dots, between dots and dashes, and between dashes and dashes, "or the dots and dashes would not be recognized."[37] That this ought not to be confused with what a visual imagination smuggles in when an ear hears a wireless transmission will have been surmised already. Thus, blindness is the price the wireless operator pays for efficient pattern recognition. Blindness grants forgetfulness, which makes keeping up with the speed of the transmission possible.[38] To summarize: wireless transmissions, regardless of transmitter, act similarly in their use of spacing to limit meaning to what can be heard; each transmission allows its listener to determine who or what is transmitting by the space or swing utilized, the wireless equivalent of Lacan's "grunt"; and each dampens the visual faculty so as to heighten the auditor's sensory capacity to a degree sufficient to perceive the transmission. Paradoxically, the wireless transmission blinds its listeners to the degree the source of the transmission demands that something be believed, to the degree the transmission qualifies as speech.

Fine-Tuning the Wireless

We are far along in our technological preamble to reading Ezra Pound's poetry, but there remains one final development before I turn to Pound's relation to the wireless. I want to begin again with what appears to be, at the very least, an alliance between war and media in World War I, with the latter providing cover for struggles between cables and the wireless. Yet media skirmishes in World War I were not limited to these two; after the development of the high-frequency alternator, wireless technology becomes the most important arena for media wars. The result will be the second great extension of the wireless, the superheterodyne circuit that transformed Marconi's humble apparatus into a device allowing for the fine-tuning of one-way transmissions across a radically larger broadcast band. In the next paragraphs, I want to describe the development of technologies to detect German high-frequency transmissions during wartime and then describe how these altered the media landscape.

The prototype of the fine-tuning dial began as an attempt by the British and Americans to detect the very high frequency waves (in the range of 500,000 to 3 million cycles) they feared the Germans were using to send messages. The problem was not simply detecting them but converting them to a lower frequency range audible to the human ear. In an ingenious discovery, a young captain in the American Expeditionary Forces, Edwin H. Armstrong, and Henry Joseph Round, a British *marconista,* devised a method in which two wireless signals, each with a different frequency, were combined to produce a third signal, the frequency of which equaled the difference between the two. Acting on R. A. Fessenden's theorem that when two different signals were mixed together they produced a third, Armstrong developed the technique known as heterodyning (from the Greek *hetero,* "other," and *dyne,* "to force together") to convert high-frequency signals.[39] For the third signal to be produced, the wireless prototype now came equipped with a wave-producing circuit able to generate a second frequency that could be heterodyned with the first. Once combined, the signal traveled through an audio amplifier to a speaker. Even the very highest frequencies were now audible through a process of heterodyning the first signal, changing it into direct current, and then amplifying the vibrations. German high-frequency transmissions, if they existed, could be detected and made audible to British and American ears.

Making sounds audible to the human ear was of course an integral part of the wireless project, from the first transmission Alfonso Marconi heard to the transatlantic transmissions of 1901. The superheterodyne represents only an advancement on the early wireless since all that was required was a second frequency with which to heterodyne the first. The U.S. Signal Corps presciently pointed this out in *Radio Pamphlet No. 40* from 1919:

> To produce a train of waves of any kind, a vibrating body is necessary. The vibrations of the body have next to be connected to a continuous medium, after which the elastic properties of the medium take care of the transmission of the waves. In the case of the electro-magnetic waves the vibrating body is an oscillatory electric charge in a circuit.[40]

The vibrating body may be the oscillating vacuum tube circuit that Armstrong and Round successfully invented to make high frequencies audible or the vibrating Marconi at Fiume emptied of life but still humming in order to fulfill its role as "continuous medium." What the superheterodyne promised and delivered was the ability to augment the inclusive capacity of the receiver. Previously unheard sounds could now be picked up with "their essential qualities of the old—its information, modulations of voice, tone of music, patterns of sound preserved."[41] In spectacular fashion, the superheterodyne opened up larger areas of the electromagnetic spectrum for transmission and reception—it became more varied than nature itself as Gottfried Benn was led to conclude since the unheard was only a superheterodyne away.[42] This in turn explains the pervasive orientation toward death the wireless continually maintained: it provided a dimension in which the unheard (and who is less heard than the dead) was combined with a second frequency that could transform it into audibility. The use of higher frequencies and shortwaves for transmission was one result, the same ones over which Ezra Pound could be heard in World War II.

As significant as the superheterodyne was in bringing higher frequencies down to a limited ear, a greater significance lay in a capacity to fine-tune transmissions. To understand what I mean, it is worth recalling wireless specs before the advent of the superheterodyne. Most sets, be they on the *Titanic* or on the Western Front, were limited to picking up strong signals, which in practice meant only signals sent through the ground or under water. Huge spark stations of tremendous power, and the Alexanderson and Goldschmidt high-frequency alternators after them, were built utilizing giant antennae so as to guarantee that signals strong enough to reach insensitive receivers would be heard. Armstrong describes the superheterodyne's radical departure from the strong signal approach in an article from 1924:

> The results obtained with this receiver were as follows. On a 3 ft. frame aerial, the factor determining the reception of the station was solely whether the signal strength was above the level of the atmospherics. The selectivity was such that stations which had never been heard before on account of interference from local stations were received without a trace of interference.[43]

The superheterodyne did what the ear was unable to do—it could distinguish between transmissions, even those being broadcast on nearly the same frequency, by a process of continual hetereodyning.[44] It was no longer simply a question of receiving a signal but of tuning into stations; weaker signals could be picked up just as easily as the stronger ones. Therefore, fine-tuning meant both a better chance of listening in to weaker stations and of reducing interference on the stronger ones, which led Joseph Goebbels to limit the broadcast band of the Nazi radio, the *Volksempfänger,* to only a small portion of the dial while pumping up the signal of state-sponsored transmissions.

The Media of the *Cantos*

Wireless (Type)Writing

With our survey of the postwar wireless landscape complete, we now turn to the case of Ezra Pound or, more generally, poetry in the "radio age" as Pound once characterized it.[45] Over the next pages, I want to examine Pound's relationship with the wireless before he began his Rome broadcasts in 1941 by first outlining the blueprints of a wireless network that includes Pound, the source of his poetry, and a typewriter. In the process, a theory of wireless systems will emerge that may account for Pound's poetic production in a remarkable canto. A systems approach to Pound's poetry is not without its difficulties (and advantages), which I attempt to meet as the chapter closes, particularly in identifying fascism as a wireless communication with its own media codes.

As I noted above, connecting Pound to a wireless network of superheterodyned fine-tuning and high-frequency alternators moves through the typewriter, so it is here that I want to begin. Pound was not the first poet to use a typewriter, but he used it more than most as a tool for composing both poetry and correspondence. Consider the following: "He first typed the letter directly, frequently crossing out words and phrases with the typewriter. Then he picked up his pen and went through his typescript, altering and adding words and phrases or occasionally a paragraph, and concluding this process of composition and correction with his signature."[46] Apparently, Pound's composition was a mix of hand and typed writing, with a predisposition to the £ symbol as a monogram: "It is interesting to note that in his typescripts Pound used the symbol £ rather than the x for his crossouts. Throughout his career he used the mark as a monogram for 'Pound.' He also used it to represent groups of poems."[47] Pound's use of the typewriter and £ to mark his poems have the following effects when compared to the handwriting that they supplemented and later replaced. Unlike handwriting, which is subject to the eye, the typewriter operates blindly, as Nietzsche's encounter with the machine demonstrates: according to Friedrich Kittler, typewriting is a blind form of writing, in the sense that the operator does not see the space of inscription but tactilely senses the layout of the keys.[48] A body learns where it must exert pressure if a letter is going to appear, allowing one to type with one's eyes closed. In addition, where handwriting flows and connects, typing implements a spatiality between letters that disrupts and disconnects them. In the curves and downward strikes of handwriting, an individuality can be discerned that is missing in the typewritten page (which the use of £ attempts to recuperate). This may account for Pound's inability to edit Part IV, "The Death by Water" of *The Waste Land*: "Bad—but cant [*sic*] attack until I get typescript."[49]

Essentially, we have the graphic difference between alphabetic printing, in which graphemes are kept separate, and alphabetic cursive writing, in which they are connected. In a letter to his daughter, Pound captures the upward stroke in an apt metaphor: "To learn to write, as when you learn tennis. Can't always play a game, must practice strokes . . . When one starts to write it is hard to fill a page. When one is older there is always SO MUCH to write."[50] Not only individuality but identifiability and coherence are diminished by the non-human strokes on a typewriter, though we do well to remember the ease with which the FBI was able to trace Pound's wartime writings to his typewriter's slightly off-center "t."[51] Finally, note how the typewriter's provision of a spatial supply of signifiers within easy reach of two hands increased the speed with which its user could produce copy. As the first monographer of the typewriter described it in 1898, "the striking of the keys follows in succession with great speed, especially when one writes with all fingers; then, one can count five to ten keyboard hits per second."[52] The adage, two hands are better than one, ideally describes the typewriter.

Like the wireless, the typewriter institutes the practice of dictation, which helps explain the ease with which the former appropriates a typewriter for its transmissions. In "Machinery and English Style" from 1902, Robert Lincoln

O'Brien states flatly that "the typewriter has given a tremendous impetus to the dictating habit, especially among business men." "The more ephemeral literary productions of the day," he adds, "are dictated, sometimes to a stenographer for transcription, and often directly to the machine." The result is "a disposition on the part of the talker to explain, as if watching the facial expression of his hearers to see how far they are following. This attitude is not lost when his audience becomes merely a ticking typewriter."[53] In Pound's case, there are a number of people and institutions operating the "ticking typewriter": his wife Dorothy and Olga Rudge (or a combination of the two); later the Foreign Broadcast Intelligence Service of the Federal Communications Commission throughout World War II; and while interned at the Pisa DTC, his daughter, Mary de Rachewiltz. Her account of Pound's dictation in the *Pisan Cantos* sets out what is required of the typist:

> I was overwhelmed by the responsibility of the typing. What if I misspelled? I remember pondering for hours over *Vai soli,* not knowing the origin of the quotation, I felt sure it must be *Mai soli,* but dared not alter it. And fountain pan?—dust to the fountain pan[.] It was only after I had detached myself from the chore and responsibility of the typing that slowly the entire passage crystallized and I could see the rose in the steel dust.[54]

De Rachewiltz's account features two important considerations. First, those who receive dictation have no guarantee that the identity of the information is the same for sender and receiver. This places her in the position of a processor of a stimulation produced by the *mai*; she ponders for hours over the difference between *mai* and *vai,* finally selecting one, not daring to alter it. The point may be obvious, but its consequences are not. When receiving dictation, she does not gain something lost by Pound, if we understand the *mai* as itself a proposal for a certain selection. The communication between Pound and his daughter requires a third term, namely, the selectivity of the information, which is itself an aspect of communication.[55] In the example, *vai* draws attention to the selection precisely because of its very selectivity; further communications begin to direct themselves to what *vai* has excluded: she ponders over the nature of the selectivity and then notes its effects. "I did not go as far as sewing fragments of the Cantos into my clothes," she writes, "but I certainly wrapped them tightly around my mind."[56] The selections Pound makes establish points of connection that generate further communication and, in the case of Pound's daughter, effect a singular wrapping. That a typewriter has already inflected these selections, limiting the noise to an alphabetic script, is a point worth pursuing.[57]

The second consideration brings us full circle in a wireless system. Initially, there appear to be two connections in the above example: that between the source of the canto and Pound, who types and handwrites; and the connection between de Rachewiltz and her typewriter. Yet these connections may be subsumed under another connection, the one between two typewriters. The speed with which one typewriter can reproduce what another has already grouped into a certain spatiality depends on the interface each uses: Pound frequently inserted his various selections in handwriting; the daughter interface is the appropriate one given the pattern-recognizing capabilities she brings. A media link is then set up between two typewriters that moves through a human interface, but how do we go about describing the nature of the partially connected media system? To qualify it as just typing confuses storing and transmitting: it is not somehow the storage capacity of Pound's brain that is measured and contained under a rubric titled "source" but rather the connection that runs between bodies and machines. The point is simply that whenever one is prompted to write, we often find ourselves in the presence of a closed circuit. Writing describes the function better but does not adequately delimit the chief characteristics: there are no wires that link bodies to their typewriters or their typewriters one to the other. Thanks to wireless connectivity, Pound's nervous pathways (of which more later) may be connected to a mechanical device, which is then routed to another typewriter through a daughter who, though lacking orthographic experience, is competent in signal recognition and speed typing. In short, she becomes a telex machine.

When the wireless began availing itself of typewriters for communication is difficult to determine. The First World War seems likely given the standardization of the component parts of typewriters that took place by war's end.[58] This meant that the wireless could utilize the same keyboard, further standardizing possible selections among available media. Both the wireless and the typewriter work in tandem, creating their own possibilities for maintenance and improvement. To take one example, Stephen Kern remarks on the scarcity of typewriters in Serbia when the government dictated its reply to the Austrian ultimatum in the summer of 1914. Eventually, the note had to be copied by hand, just making its date with history. Although a telegraph carried that message, when the United States issued its terms to Germany in 1918, the wireless station at New Brunswick, New Jersey, received President Wilson's now typewritten dictation. Here too the war appears to have accelerated a wireless coupling with the typewriter, leading some like Paul Virilio to argue for a history of wireless connectability among storage media, the opening gambit of which is World War I. However enticing wartime coupling may be, it ignores the process whereby wireless operators learned to approximate typing machines by becoming anonymous writers. Training *marconisti* to transcribe in capital block letters when listening in began much earlier than the war; it seems to have been a mainstay of wireless transmission almost from the time of the first transatlantic "S." The move from an anonymous blind writing in block letters, whose source was

weak spark transmissions, to an anonymous blind writing in typewritten letters, whose source was stronger, alternator-generated signals, was a short one.

Put differently, the wireless acts as a multiplier of uncertainty. It changes
typed or block letters into dots and dashes, two minimal signifiers that in turn
are changed into electric current to be transmitted through the air. The result is
a separation of information from source. The typewriter mechanized the function of the human interface on the wireless, spatializing a general stream of
data, readying it for transmission into a binary of dots and dashes. A dictated,
typed message presents fewer problems for the operator: spaced-out letters are
more easily transposed into dots and dashes than handwriting. Information
and uncertainty increase when the typewriter is hooked up, and a separation
ensues between the source of the communication, be it writer or poet, voice or
hand, and information, understood not as meaning or content but as the set of
possible selections.

Wireless(ed) Cantos

In 1924, Pound registers the impact of the wireless in a letter to his father: "As
to Cantos 18–19, there ain't no key. Simplest parallel I can give is radio where
you can tell whose [sic] talking by the noise they make."[59] But even if Pound
had failed to equate the switching operations of the *Cantos* with the radio, linking recognition of speakers with "the noise they make," it would still be necessary to account for the appearance of voice registration in these cantos. Pound's
reading instructions recall the inverse relation between noise and signal and
point to a crucial effect of wireless communication—in literary terms, the difficulty in locating the source of transmissions wirelessed to Pound becomes
the content of the canto. This is because one medium, poetry, begins registering another medium, the voices Pound picks up and puts down in verse. These
wireless cantos commemorate their debt to the advanced wireless circuitry of
the day in nonlocalizable sources, missing addresses, and frequent switching
operations. Within the structure of *A Draft of XXX Cantos,* they leap ahead of
the (relatively) simple transcriptions and translations of sources that constitute
Cantos 1–17 precisely because they register more noise.

Initially this does not appear to be the case. Canto 18 opens with an extended transcription from Marco Polo's diaries concerning Kublai Khan, a fact
we discover in line 26, "—thus Messire Polo; prison at Genoa—." Any reader
familiar with Pound will recognize the translation/transcription as typical of
A Draft of XXX Cantos (see the Malatesta Cantos, especially Canto 8, in which
Pound translates letters and archives and reduces them to what he calls "fragments"). The next lines, from "Of the Emperor" (line 26) to "He owned a fair
chunk of Humbers" (line 39), I want to hold in reserve and direct my attention
to what follows:

> " Peace! Pieyce!!" said Mr. Giddings,
> " Universal? Not while yew got tew billions ov money,"
> Said Mr. Giddings, " invested in the man-u-facture
> " Of war machinery. Haow I sold it to Russia—
> " Well we tuk 'em a new torpedo-boat,
> " And it was all electric, run it all from a
> " Little bit uv a keyboard, about like the size ov
> " A typewriter, and the prince came aboard,
> " An' we sez wud yew like to run her?
>
> (ll. 40–48)[60]

Not only do the lines differ markedly from the canto's previous thirty-nine lines, but they are distinguished from the previous seventeen cantos to the degree they register a voice, in this case that of a certain arms manufacturer, Mr. Giddings. A glance at Carroll Terrell's *A Companion Guide to the Cantos* confirms that Pound has not transposed them from a book, which is to say their content is not simply that of writing, be it a fragment from Marco Polo's diary or Sigismundo's accounting records, but rather of a voice and its modulations. Proof may be found in the onomatopoeia that litter the speech, the abbreviations used to mark the voice (the "sez," "wud," and "yew" that are commonplace in the *Cantos*), and the sentence fragments that constitute Giddings's speech ("Haow I sold it to Russia" and "And it was all electric").[61] They are qualitatively different from, say, "'Bring 'em to the main shack,' said Baldy" or " Here! this is what we took out of you" (to take two early examples of voice transcription from Canto 12), for the sheer length of the registration and the effect they maintain. The typographical manipulation that a typewriter makes possible is allied with voice registration to isolate fragments of speech by utilizing the space in and between lines of poetry. One notes the quotation marks that mark each line as possibly discontinuous from the last, and the space between quotation and quotation mark that follows each successive line. This works against continuity, isolating punctuation marks from letters, handicapping an easy fluency in reading across the line. The effect is to slow the reader down, of which the opening's one and then two exclamation points of "Peace!" and "Pieyce!!" are emblematic; skipping over them is nearly impossible.

Therefore, we can date to the year that Pound began picking up his first voice transmission; it was 1924, long before he began listening illegally to shortwave BBC transmissions in Fascist Italy. We can also name the process of picking up and writing down a voice—it is what Pound defined as "a sort of energy, something more or less like electricity or radioactivity, a force transfusing, wielding, unifying."[62] In other words, poetry as data processing. Written archives alone, the nineteenth-century storage media of the kind Pound mentions in regard to T. E. Hulme in Canto 16 ("With a lot of books from the library"), is extended two cantos later to include the voice an ear has picked up. That we are in the realm of technology is demonstrated again by the healthy future voice registration has in the rest of *A Draft of XXX Cantos*. It becomes clear the further we read that any voice that meets "this sort of energy, something like electricity" can be turned into a work of poetry, which is to say that a voice registered in poetry shares much with the wireless since it writes down the particulars of a voice that few readers will recognize. But perhaps we need to link more properly the data processing that goes under the name of the *Cantos* with wireless technology.

The Switches of Quotation

Canto 18 does not register just Giddings's voice but a number of voices and then successfully switches between them. Consider the following samples:

> And Mr. Oige was very choleric in a first-class
> From Nice to Paris, he said: "Danger!
> " Now a sailor's life is a life of danger,
> " But a mine, why every stick of it is numbered,
> " And one time we missed one, and there was
> " Three hundred men killed in the 'splosion."
>
> (ll. 71–76)

> So I said to the old quaker Hamish,
> I said: " I am interested." And he went putty colour
> And said: "He don't advertise. No, I don't think
> You will learn much." That was when I asked
>
> (ll. 87–90)

> " Qu'est-ce qu'on pense . . .? " I said: " On don't pense.
> " They're solid bone. You can amputate from just above
> The medulla, and it won't alter the life in that island."
> But he continued, "Mais, qu'ᴇsᴛ-ᴄᴇ qu'ᴏɴ pense,
>
> (ll. 94–97)

If the Mr. Giddings selection assumed something like competition with straightforward voice transmission, then these three samples presuppose technical competition not with books or with one voice but with a transmission media capable of conducting switching operations from one voice transmission to another. There is no archive that rhetorical figures deploy, as in the Odysseus fragments of Cantos 1, 2, and 3, but rather the switching between voices processed by a transfusing poetry. Where the Mr. Giddings section wore its voice registration on its sleeve in onomatopoeia and isolated bits of just overheard bravado, the three selections throttle any means of identifying their sources outside of announcing them. They lack the detail of Mr. Giddings, possibly because they jump from one voice to another.

In other words, the bandwidth spreads as Pound collects data. The effects may be gauged in the lack of a destination for the transmissions since they are not addressed to a reader who can confuse sensory input with the data itself, hallucinating a meaning out of what are in fact only printed letters. The small capitals of line 97, "Mais, qu'EST-CE qu'ON pense," confirm this by drawing attention at once to the mediated quality of what we are reading (and not speaking since it is unclear how one would voice them). In the first example, a "very choleric" Mr. Oige is recorded in pieces by a media system that readies the conversation for later transmission. But something happens between storing and writing it down: when the Real moves through a media linkup, noise and information are the result. In the above selections, noise is measured in what we might describe, along with legions of Pound's critics, as a lack of coherence. To take another example, line 73 ("Now a sailor's life is a life of danger") and the next ("But a mine, why every stick of it is numbered") are seemingly at odds; the conjunction does not logically follow from the previous line, indicating that perhaps a condensation has occurred. Indeed, the more we read, the less coherent the selections become. The "I" of lines 87 to 101, who addresses the old quaker Hamish with "I am interested" is met with "He don't advertise," an incoherent response given what "I am interested" fails to qualify. Doubtless, there are the repeated references to Zenos Metevsky (ll. 80, 91, and 99), which serve, to borrow a Poundian metaphor, as a method for shoring and shelving fragments. They function throughout the canto as a rubric with which to group the various voice transmissions, the call letters of usury and the merchants of war. But they do not provide the reader with all the destinations for the transcribed voices; this is confirmed in line 100 when a registering "I" tells a French-speaking no one in particular that "They ain't heard his name yet" when referring to Metevsky. As in the Giddings selection, the typography works along the same lines, with the repeated use of quotation marks effecting a separation between voice and its registered, mediated nature. The "No, I don't think / You will learn much" of lines 89–90 is the only exception to what appears to be the case when Pound registers voices: the relay running from the canto's storage medium to the typewriter can only carry a certain number of words, which it transposes across a single line of poetry and no more. The quotation marks that begin the next line mark both the words not selected and the compression of what will follow.

Undoubtedly, this is of a part with Pound's aesthetics articulated as early as *Blast* and later in *How to Read* and *ABC of Reading*. Yet it is not enough to say that Canto 18 is incoherent in spots due to the peculiarities of Pound or the nature of what he is registering.[63] Pound's reliance on typography and the spacing a typewriter provides does more readily account for what appears before the reader and the heightened condensation of the rest of the canto (see, for instance, the "wd." of line 109: "The Germans wd. send him up boilers, but they'd"). However, what agency and an encounter with the typewriter cannot account for are the repeated violations of linear time, the unfeasibility of lines 26 to 32, in which "Of an Emperor" may denote Kublai Khan, Napoleon, or the arms magnate Metevsky. The movement between presence and absence is best captured in line 36, when Metevsky, "died and buried," watches his own funeral while sipping a coffee. Where agency and typewritten effects fail, the switching operations of the wireless more fully account for these shifts in time. When poetry becomes a site in which voice is registered in writing, it separates the source from the event of transmission. When poetry effectively switches among voice transmissions, it speeds up the transmission and transgresses its customary boundaries.[64] Voices begin to move across frequencies and time; voices like Napoleon's or Metevsky's are no longer anchored to a line of poetry but migrate across lines and pages. It now becomes possible to manipulate the elements of what went under the rubric of poetic creation or inspiration one hundred years before so that not only is it unclear who is speaking but when the registered speech might have occurred.

In the first quarter of this century, two transmission media increased their transmission speeds, began picking up and writing down voices, and developed a capacity to switch frequencies. The result was the same: an increase in the number of frequencies and with it greater uncertainty regarding source attribution. Uncertainty due to a greater number of frequencies allowed for time displacement so that the time of a transmission could only be localized with great difficulty. How many years separate a twelve-year-old Napoleon/Metevsky/Kublai Khan from one another? Any attempt to limit the attribution to one speaker and one frequency, and hence to fix the time of the speaking event, risks ignoring the near frequencies that may be migrating across bands. The point is simply that time and frequency displacement in wireless transmissions, be they strictly on the wireless or in poetry, can never be measured exactly.[65] Consider too that a listening Pound and a functioning wireless produce a dispersion of voices; each can transmit heard and unheard voices. The wireless accomplishes this by heterodyning the signal—Pound does so by increasing the switching operations to such a degree that, in the frequency displacement represented by fragments and bits of broken conversation, what follows in omitted words and the white of the printed page is never just the trailing voice of Mr. Ogie or Mr. Giddings but the difference between them and among all the voices registered in the canto. Typographical space becomes the background on which the unheard is heterodyned into the next line, a tone imprinted on both punctuation and space. The consequences are startling. The canto consumes its media sources by transforming books, voice transmissions, and overheard banter into a stream of data that moves through the supply of signifiers a typewriter provides. The resulting series of letters and spaces represent the maximized dispatch possibilities of the wireless interface.

If the content of a medium is another medium as McLuhan and Derrida argue, it behooves us to identify the media the canto has successfully connected.[66] The sources for Canto 18 turn out to be either sequences of writing (the Marco Polo section and the King Menelik tractor anecdote that concludes the canto) or voices. In both cases, the written word and voice are transposed by hands that type into a sequence of lines. These inputs, writing and voice, are themselves transmission media, which the canto distributes across the typed page. In this regard, one recalls Marinetti's revelry on the biplane and the function of the wireless for a Futurist aesthetic. It was not just the product of a Futurist love of heights that made *parole-in-libertà* possible, but the wireless connection running between the propeller and Marinetti's ears joined to a writing hand. The transmission quality of Marinetti's wireless imagination, its high and low definition, depended on the storage medium and the channel connecting it to his writing hand. The earlier example of Everett Gordon's wireless operators is instructive. Most were urged to transmit at no more than eight words a minute, given the slow connection speed between ear and hand. With the advent of voice transmission, wireless definition improved due to the greater variety of frequencies the human voice contains. It was precisely the greater range of the human voice that allowed the wireless to limit its definition and still be comprehensible, as speech-scrambling tests at Bell Laboratories demonstrated:

> One test, for example, eliminated (by electrical means) all sounds below 100 c.p.s. [cycles per second] in a series of nonsense syllables. Subjects missed less than 10 per cent of the syllables, and in running speech probably would have lost nothing. . . . The excess of detail defends speech against the noise and accidental of ordinary activities by ensuring that, even if one component is eroded, the others will sustain the message.[67]

The result is another theorem of the wireless transmission: the transmission limits of the storage medium in use dictate what we take to be the meaning of the transmission.[68]

In Canto 18, the passages that contain more detail (onomatopoeia and altered orthography) are precisely the ones that register voices and the noise of the transmission. Not coincidentally, those sections for which the source is a written archive exhibit less noise to the extent they are more compatible with a data processing still anchored in writing. Essentially, we are speaking about two

partially connected media systems: on the one hand, writing or, more specifically, typed poetry that is only capable of storing itself and, on the other hand,
a wireless system able to connect a variety of voices (stored on phonographs or
later, in the case of Pound's radio transmissions, on disks) to listening ears. But
the connected media are formatted differently—writing is incapable of storing
voice as Derrida has so persistently pointed out.[69] This accounts for both the
detail of Mr. Giddings's speech and the noise of the transmission, registered in
the section's fragments, missing predicates, and spatial gaps. Canto 18, it bears
repeating, measures the incompatibility between two media systems that are
coupled by writing. Perhaps Pound had this in mind, not only when he allied
voice recognition with the amount of noise the registered voices evince, but also
when he armed his poetry with a noise-registering capacity.

Finally, consider generally what writing registers. Before the advent of this
century's technical media, what was received and emitted fell to the monopoly
of the printed word: "Whatever they [poets, philosophers] were emitting or receiving was writing. And since whatever exists depends on what can be posted,
the bodies themselves were submitted to the regime of the symbolic."[70] When
switching operations begin, poetry's transmission limits remain tied to the
storage medium of writing. Formulated differently, writing couples a wireless
medium to poetry with the effects I previously noted: time axis manipulation
and voice migration that lend a hallucinatory quality to the proceedings. But
Canto 18 transmits only what writing stores. On the one hand, speedier transmissions and with them the higher definition of wireless voice transmissions
short-circuit meaning. On the other, storage limits restrict what can be sent
across the page to twenty-six letters.

Pound's Wireless Aesthetic

If Canto 18 measures the distance between two media, we might well ask how
Pound functions in the coupling, where he is located, and what role he plays
in poetry that bears his signature. In other words, is Pound the source of his
poetry or is he just another piece of information on the wireless interface in
use? Pound senses the significance of the question, and if we take him at his
word (not always an easy task), he more often than not decides for the latter.
While much is inevitably lost in the cursory summary of Pound's technological
aesthetic that follows, I hope to show how indebted Pound's aesthetic is to the
wireless. Three quotes set the parameters for the discussion. "The permanent
property," Pound writes in "The Serious Artist," his most detailed examination
of the question, "the property given to the race at large is precisely these data of
the serious scientist and of the serious artist . . .; of the serious artist, as touching the nature of man, of individuals."[71] Second, there is the famous opening

to "The Teacher's Mission": "Artists are the antennae of the race."[72] And finally, we have Pound's exposition on chaos in "Dr. Williams' Position": "Art very possibly *ought* to be the supreme achievement, the 'accomplished'; but there is the other satisfactory effect, that of a man hurling himself at an indomitable chaos, and yanking and hauling as much of it as possible into some sort of order (or beauty), aware of it both as chaos and potential."[73] Taken together, the quotes identify the function of the poet, the means by which he accesses the transmission, and the noise that occurs whenever a symbolic transcription of the Real is attempted. Below I want to examine each in turn.

In Pound's estimation, the artist, frequently substituted with "poet" in the essay, enjoys a privileged relationship with the "race" to the degree he gathers data that "touches on man." To be sure we have not missed the point, Pound repeats it: data concern "the inner nature and conditions of man."[74] The use of data to describe what formerly might have been termed content was admittedly standard for the period in question—literature was busy cannibalizing science's metaphors, making assessments of the impact of science and technology on literature difficult.[75] This was perhaps truer for Pound than for others, given his penchant for the apt technological metaphor. Yet my reading of Canto 18 and its debt to the wireless shows that we are not simply in the realm of scientifically inflected terminology. The canto attempts to store data (voice), to process it (in typed letters and heterodyned spaces), and to transmit it (the accelerated transmission speeds of voice that break up writing). There is no poet on one side successfully modeling his poetic language on a wireless technology but, rather, one medium in competition with another. When poetry becomes data retrieval, the poet's task is to access the Real.

There are of course a myriad of ways of accessing it. D'Annunzio picked up the Austrian marconigram traveling through a boundless, timeless ether and, in the process, was turned on to the intoxicating effects of wireless technology. For Marinetti, autism was allied with speed to create the conditions under which a writing hand could begin writing down transmissions wirelessed from a propeller. Intoxication, speed, and sensory aphasia all present points of contact with the Real. In each case, the wireless sets bodies in motion, inching them closer to the transmission's source, distributing senders and receivers in a network of relays that could send words and sounds. Traditional notions of prosthesis fail to describe these effects of wireless technology since it is not only a question of extending a body's possibilities or even that the wireless was articulating "an altogether new set of possibilities" for the body.[76] The wireless enlists bodies in its ever-widening network of operations, but there is no localized body on one side receiving malevolent effects from the wireless on the other. I am not denying that wireless effects can be measured on the body, but when attention is given to the operations of a media linkup, such effects will appear as the result of a coupling between the wireless and poetry: quantifying information shifts our attention away from agency and prosthetics to couplings.

With this preface complete, Pound's equivalence among artists, antenna, and the wireless becomes more manageable. When he writes, "artists are the antennae of the race," he identifies the location of the artist and poet when data processing becomes their chief obligation. It is to listen in, to turn one's ear to the Real, and to pick up transmissions for the benefit of the race. These are not simply metaphors—Pound was listening in to transmissions before and after he composed the essay—nor do they merely indicate a wireless installation in Pound's body that extends his perceptional capacities. Pound offers job descriptions of poets and artists when poetry attempts to compete with wireless technology by transmitting voices; they are the interface between a real data flow and its symbolic registration. The poet sits at his typewriter and attempts to listen in to the Real; the data flow passes between him and his spacing apparatus. In such a scheme, antennae index a poet's function in a wireless network. A similar connection was operating in Marinetti's "Technical Manifesto," but where wireless transmissions were routed through one hand, the voice transmissions of Canto 18 are routed through a typewriter that readies them for printing.

The third quote confirms the relationship the poet enjoys with the Real and the symbolic. On first reading, Pound's image seems paradoxical to the point of incomprehensibility. A man hurls himself "at an indomitable chaos" and, "yanking and hauling," puts it into some kind of order or beauty. The next sentence immediately allies chaos with potential through the medium of consciousness, for man is "aware of it." That is the key: Pound is equating patterns of recognition with aesthetics. Before its ordering in the "yanking and hauling," chaos is recognized as both chaos and potential. Reading backward then, the object of chaos's potential is order: chaos's potential for ordering is also its potential for beauty. In the passage the Beautiful helps the imagination organize facts, but more important it is the rock and the hard place between which an artist, intent on "accomplishing" art, finds himself. The rock is the indomitable chaos; the hard place is order maintaining its potential as chaos; and in between is the man who hauls as much as possible out of the chaos. Pound places the poet at the intersection of the Real and its symbolic ordering, which is typical of Pound's aesthetics.[77]

Even so, objections to this reading may be raised regarding agency and the verbs with which Pound describes what a man must do to achieve art. He appears to be naming the process by which a goal is actively pursued. We might agree were it not for what is missing from Pound's aesthetic: there is no Muse ("O my Clio! / Then the telephone didn't work for a week" [Canto 19, l. 85]), no inspiration of the Romantic sort, only order and chaos and art refashioned to coincide with media conditions. In addition, we need to consider the command to maximize the capacity of what a man may move from chaos into order. One hundred years before, it was not a question of yanking and hauling "as much as possible" of chaos into order since the real flowed only through writing. Pound indirectly recognizes writing's limitations when he writes: "Before the experimental method, when men had hardly more than words as a means for transmission of thought, they took a great deal more care in defining them."[78] His formulation respects the new media conditions under which art functions. When the symbolic is recognized as holding on to its potential for chaos, that is, as a selection from the Real, man is found carrying out an inhuman command to maximize capacity.

A question follows. Who or what has sent Pound the command to yank and haul "as much as possible?" As we discovered in the last chapter, commands to listen and write proliferate in the early part of this century, first issued by psychophysicists, like Sergi, who were intent on determining the transmission and storage limits of their patients. The wireless takes over the reins when Marinetti initiated his *parole in libertà*. As wireless transmission media come to dominate the landscape, one senses the same process in Pound: the wireless issues a command to write that which was formerly addressed by a slew of positivists. Writing on Gottfried Benn and radio, Friedrich Kittler might be speaking of Pound as well: "In Benn's early work, the data were taken from everyday talk, the addresses in an associative brain, and the commands from a doctor, who with his whip, called for words or texts. The poetics of radio transferred a miniature of this command function to a consumer who could turn the transmissions on and off."[79] Kittler focuses our attention on Benn's brain when describing the effects of a radio transmission. If we are to determine the source of the command to full capacity in Pound, we would do well to examine him in like fashion. Fortunately, we are preceded by an impressive number of psychiatrists at St. Elizabeth's Hospital and in Italy, who from 1946 to 1968 listened to Pound on a weekly basis. They may help us identify the source of the command to write.

"Speaking as He Writes"

In an undated examination of Ezra Pound, perhaps from 1968, Dr. Cornelio Fazio, of the Clinic for Nervous and Mental Disorders at the University of Genoa, identifies the source for Pound's poetry. Pound had previously sought help from Fazio in 1966, who at the time diagnosed him as suffering from "an extreme autistic situation." "On the other hand, the autistic personality traits were always evident, perhaps even increased, owing to the failure of expansive

hypercompensation mechanisms: the pattern of thinking seemed to follow particular inner associations, tending to meet internal needs, in disregard of objec-
tive reality." The doctor goes on to describe the nature of Pound's autism:

> It seemed as if the personality of the patient had always been on the autis-
> tic side, with a prevailing phantastic attitude and insufficient contact with
> reality, inadequacy of associative structures of thinking, of pressure of a
> particular type of association, which could be considered hyperinclusive and
> "à côté," so that a psychotic-like situation came out ("borderline patient"),
> permitting however, and perhaps, encouraging poetic activity.[80]

Fazio's insight into the relationship between poetic activity and the inadequacy
of Pound's associative structures of thinking indicates a connection between
deficiencies and poetic production. But what exactly does Pound's physician
have in mind when he writes that a "particular kind of pressure" keeps Pound's
brain associating in a certain way? As I hope to show, the answer lies less in a
localized area of the brain that functions improperly than in a partial system
of connected media that keeps Pound associating. This represents another advance on the wireless in Marinetti. The media in question we already know:
Pound's brain as the primary storage medium; a Corona typewriter to provide
a ready supply of letters and spaces when voice transmissions begin; and the
wireless that commands that Pound type (and later speak), as it connects his
body to a device that can immediately prepare copy for publication.

Still, some second opinions are in order concerning the autistic diagnosis.
In a case history taken down on January 24, 1946, Dr. Kavka, the examining
psychologist at St. Elizabeth's, notes the rapidity of Pound's speech. "The patient spoke rapidly, with a faint suggestion of an 'English' accent, and modu-
lated his voice frequently, from a barely audible whisper to shouting 'at the top of
his lungs.'. . . At times, he would soliloquize in a manner typically found in his
poetry—apparently disconnected thoughts and ideas striving together in a pattern."[81] Three months later, another psychologist from St. Elizabeth's observed:

> At times his speech is fragmentary, although telegraphic in style, resembling
> the cryptic letters he writes. In fact, his present style of speech and writing
> resembles his poems and other artistic productions. He is apparently a true
> Symbolist, who compresses a large volume of words and concepts into a
> brief expression. . . . Often when asked to amplify an obscure remark, he will
> reply that his ideographic processes move rapidly but are not distorted or
> obtunded.[82]

When prompted by his doctors, Pound compresses "a large volume of words"
into a brief expression; he frequently "modulates" his voice, and his "apparently
disconnected thoughts" are both "telegraphic in style" and similar to the "cryp-

tic letters he writes." In other words, Pound speaks as he writes, the spoken par-
alleling his condensation on the written page. In this regard, Pound's letter to
James Laughlin concerning condensation in Cantos 52/71 (Pound's notation)
is instructive. Pound writes that "the new set is not incomprehensible. Nobody
can summarize what is already condensed to the absolute limit."[83] Put differently, the competition with the wireless that prompted Pound to pick up voices
first in 1924 was taken over by his doctors in 1946.[84]

In fact, the more one reads of the hundreds of pages of notes and diagnoses, the more Fazio's deficiency in associational thinking begins to resemble
a complex information system. Pound's doctors are on the receiving end of a
dictation, occupying the place previously held by a typing daughter or simply typing hands. If the Real that a poet transposes into the symbolic moves
through media, a typewriter or a psychologist will do equally well in taking
down copy. Naturally, the doctors receive Pound's transmission since they were
the ones prompting him to maximize his dispatch capacity in the first place.
But in both instances, the connection moves through Pound's body. To understand what I mean, consider the number of channels that connect Pound's wireless voice transmissions to the St. Elizabeth's staff. Pound speaks as he writes,
utilizing telegraphic style, condensation, and amplification of voice. Increased
noise is registered in his "apparently disconnected thoughts" and the "cryptic
letters" his doctors have no doubt perused, not to mention the gramophone
Pound's chief doctor at St. Elizabeth's used to play back the poet at his leisure.[85]
To the extent they receive what he sends, Pound is commissioned to cover all
the modalities of technological media. However, the sheer breadth of coverage
changes nothing. In each case, couplings with doctors or the typewriter exhaust
the limits of what the symbolic can transpose from the Real. The material each
takes down is intended for storage and transmission, be it on paper or on a
long-playing record.

There is a further point pursuant to noise. When asked to "amplify an obscure remark," Pound denies that his ideographic processes distort the Real. To
answer in this way requires him to make no distinction between the written
and spoken under the rubric of the ideograph. This is not happenstance: as the
earlier section devoted to voice recognition on the wireless demonstrated, the
dimension of speech in a wireless transmission resides in a desire on the part
of the sender for recognition. Odysseus's men transformed into pigs attempt
to grunt their way to recognition despite the noise levels inherent in such a
transmission. When a poet is busy exhausting his capacity to "yank" as much
as possible out of the Real, he faces the same problem. Fragments lack addresses
because processing information results in data about the media link while failing to identify who is speaking (given the conditions under which the transmission has taken place). Perhaps this is what Pound had in mind when he wrote in

Guide to Kulchur in 1938: "There is no ownership in most of my statements and I cannot interrupt every sentence or paragraph to attribute authorship to each pair of words, especially as there is seldom an a priori claim even to the phrase or half phrase."[86]

The received message, taken down by a psychologist or a typewriter, is addressed to everyone and no one and occurs by way of a prompting issued by a doctor or a wireless capable of picking up voice transmissions. The received signals display greater information to the degree they carry more noise, which is not unexpected considering the prior reading of Marinetti and how wireless transmissions exhibited their noise levels in missing adverbs and a new orthography. For both Marinetti and Pound, the more uncertain we are who is speaking, the greater choice we have in selecting a source. Delimiting the source for Marinetti's wireless writing was relatively easy, however, since the conditions of media competition under which he wrote did not include switching operations and stronger signals. Pound's commitment to the non-noise of his ideographic processes appears incorrect then because it confuses uncertainty that comes about "by virtue of freedom of choice on the part of the sender" with "uncertainty that arises because of errors or because of the influence of noise."[87]

A relation that runs between a brain and a dictation medium issues commands to write down or speak exhaustively. It creates noise, which becomes the transmission's content. When such is the case, then listening on the wireless or tracking switching operations in the *Cantos* is finally an exercise in voice recognition. Paraphrasing Lacan, the wirelessed voice seeks recognition that it is still human and not just another ghost addressing the dead. What is left to Pound in the ever-expanding reach of the wireless then is the meager possibility of a pattern in the fragments that can somehow resist the entropy of the wireless and the information (and uncertainty) it produces. Canto 110's appropriation of Eliot's "These fragments I have spelt into my ruins" inevitably comes to mind:

> From time's wreckage shored,
>> these fragments shored against ruin.
>> (l. 781)

where entropy is measured in fragments "shored" (and stored) against time.

Fascist Media Codes

Pound's exposition on the properties of his "ideographic processes" indirectly raises the question that haunts the wireless in each of its manifestations: what does the relation between user and wireless indicate generally about fascism and technology? The question is germane when recalling Pound's unwavering

commitment to fascism and the importance he gave technology when accounting for his poetic production. After reading Pound's voice transmissions in one
of the *Cantos* and recognizing the systemic role writing them down has in a
system of partially connected media, we might rephrase the question this way:
how can we describe fascism in terms of systems of connected media and, if so,
what media constitute it? The breadth of such a project is not within the reach
of the present chapter, for it requires a theory that can account for the interpenetration of various media, including cinema and the gramophone, with the
notoriously elusive psychic systems of individuals. Yet, if we limit our perspective to the terms previously used in regard to D'Annunzio and Marinetti, we
may indicate with some precision the form such a project may take.

What does Pound's denial of noise to his wireless processes assume? Pound
posits a connection between reader and poet beyond the short-circuits of meaning theorized by Marinetti and put into practice by Il Duce. There is no intermediary between the poet and reader, which restates the position that poets are
the antennae of the race while recalling an electrified past that grew up around
the wireless. As important as rhetoric is for fixing the precincts of condensation
and speed in fascism and poetry, the present chapter indicates another possibility. In a system of connected media, Pound's poetry is the result of a series
of media couplings that prompt a poet to register information. A typewriter
joins a body to a wireless media system, which by 1924 was capable not only of
picking up voices but of switching between them. Canto 18 registers the new
media landscape because it too finally is a medium and media serve to bracket
other media.[88] One may well ask why poetry seems especially given to registering its competition with wireless media while cinema and the gramophone
seem less inclined. First, unlike the latter two, the wireless is a transmission
medium capable of sending writing between two points. The form of writing
was in a sense ready-to-order for successfully coupling the wireless with poetry:
by confirming the medium of poetry, writing also confirmed poetry's potential
for other couplings. Second, as the wireless reconfigured the media landscape,
poetry was no longer just competing with a storage medium (books) but with
a transmission medium that could connect storage media (books, brains, and
gramophones) with ears and hands. The canto puts into operation a variety of
coping strategies to keep up with the growing speed and importance of other
transmission media.

Naturally, as our perspective on Pound's poetry changes to reflect these
developments, so too will the importance we give fascism in accounting for
Pound's poetic production. A statement such as "fascism appeals to Pound,
by his own testimony, precisely because it dispels verbal complexities, since it
means 'at the start direct action, cut the cackle, if a man is a mere s.o.b. don't

argue'" places fascism before media and then proceeds to find it lurking behind Pound all along.[89] Moreover, it fails to account for the effects of one medium on another since the category of fascism alone is incapable of dispelling verbal complexities; that role falls to transmission media of the written and wireless variety that had been agitating for greater condensation twenty years earlier. The power of Pound's poetry to transform voices into writing is an index of the power of the wireless to decrease the distance between individuals and Il Duce. This represents an extension of its capacity at Fiume. Not only does it bend ears as at Fiume to its apocalyptic transmissions, shortening the distance between D'Annunzio's mouth and his listener's ears, but now it extends transmissions over greater distances while enabling switching operations among a number of voices.

I am not speaking of the radio in its 1930s format nor denying that radio broke down "demarcations of private, public, individual and community which were so central an element of the fascist grasp of power."[90] Rather, it would seem that there is a maximization of dispatch modalities across a broader spectrum, addressing and mobilizing an ever greater number of people for war. This is what Reich minister of armaments Albert Speer called at his Nuremberg trial the product of technical media: "The telephone, the teleprinter and the wireless made it possible for orders from the highest levels to be given direct to the lowest levels." The lack of mediation between the supreme leadership and the lowest levels of the command structure meant that orders "were carried out uncritically."[91] To cope with the greater strategic ranges of a general military mobilization, poetry maximizes the modes of transmission available to it, making it possible for orders to be carried out uncritically.

Delmore Schwarz senses the first half of the equation when reviewing the Middle Cantos: "It is not only that some great modes of poetry—direct statement, description, speech, and the movement of the poem itself—have been given fresh kinds of rhythm, but that, above all and extreme as the claim may seem, our capacity to *hear* words, lines, and phrases has been increased by the Cantos."[92] Although Schwarz misses the media competition that motivates the use of ever more modes to reach the reader, he does recognize something new in the *Cantos*. An increased capacity to "hear words" is the result of utilizing fresh rhythms and all the modes of dispatch available, which, as my reading of D'Annunzio demonstrated, places the *Cantos* in the neighborhood of an oracle demanding sacrifices. And no ears are ever opened without the required sacrifice, as Pound records in the first Canto:

> For sacrifice, heaping the pyres with goods,
> A sheep to Tiresias only, black and a bell-sheep.

Dark blood flowed in the fosse,
Souls out of Erebus, cadaverous dead, of brides
(Canto 1, 26–29)

Writing in 1960, W. D. Snodgrass makes out the second half of the equation, capturing the nexus among sacrifice, interpellation, and poetry in what he calls "flash-card" verse:

When they [poets] achieve power, they must face facts once again—including the facts of their own intention. And if I am to judge from that area where Pound *has* power—poetry—his insistent domination of my mind grows finally very binding and distressing. I think we do well to suspect that domination regardless of stated intentions. I find the urge to enlist under any benevolent Throne or State (existent or nonexistent) thoroughly resistible; I am continually reminded that it was the Army indoctrination course which most successfully used the flash-card, teaching us to select those we would kill with a minimum of thought beforehand.[93]

Snodgrass imputes the "power" and "domination" of Pound's poetry not to a simple notion of agency but to the effects of a medium. Granted he has the "Thrones" section of the *Cantos* in front of him, in which hieroglyphics and bursts of ever more fragmented verse increasingly become the norm as voice transmissions fade out: attempts to match the absoluteness of speed-of-light transmission strategies occasion the one word, the one cipher. In Canto 18, maximized dispatch modes are restricted to direct statement, speech and description, but the effects differ only in degree.

What we have is the possibility that fascism, as a mode of wireless transmission, maximizes dispatches to enlist bodies for warfare. But surely, the reader will wonder, are we not reading too far ahead into Pound's politics when we link the canto via wireless media with Pound's political thought that was still ten years away?[94] Or more significantly, how does substituting the totalizing tendencies of Pound's politics with an amorphous concept of wireless transmission account for Pound's anti-Semitism and his heavy-handed dichotomization between friend and foe? In his brilliant history of World War I English literature, Paul Fussell offers another possibility. Describing the "special sensitivity to division" after World War I, he cites the famous injunction on the title page of *Howard's End,* published four years before the war: "Only connect." "To become enthusiastic about connecting," he writes, "it is first necessary to perceive things as regrettably disjoined if not actively opposed and polarized."[95] Of course, Fussell is interested in manifestations of an adversarial habit of mind in a weltanschauung that grew out of the war, in England specifically, and not in the workings of a partially connected media system registered in Pound's poetry. Yet his reading reminds us that transmission media are constantly in the business of creating binaries so that fascism and transmission media are collaborators in their efforts to (dis)connect. As the wireless increases its speed and capacity to carry different voices, it begins drawing ever starker lines between friend and foe. As Pound's poetry attempts the same, the anti-Semitic references and gross dichotomies increase, reaching a technological and adversarial ecstasy in his radio broadcasts on Radio Rome. At the risk of reducing too much, the wireless appears bent on programming Pound for a role within a particular wireless subsystem known as fascism.

"Ezra Pound Callin'"

If I have demonstrated anything in this chapter, I hope it is the significance of writing tools and transmitting devices for Pound's poetic output in the first quarter of this century. Such an interpretation has its advantages, especially when we begin listening in to Pound's Rome broadcasts. Much of course has been made of Pound's use of the microphone, not least by Pound himself ("Give me a bed, a bowl and a microphone"), but less of the emerging media network that made possible the recording and transmission of his voice.[96] As we turn to these other, more famous wireless transmissions, the early voice transmissions captured in Canto 18 will provide the coordinates for locating once again Pound's debt to wireless media. Therefore, my final object of study will be Pound's Rome broadcasts, the media technologies that conditioned them, and the noise that, paraphrasing an early critic of the *Cantos,* Pound's madness keeps throwing out.[97] With some effort we may be able to identify noise on the radio. My task will be aided immeasurably by Rudolf Arnheim's guide to voice transmissions, *Radio.*

5

POUND'S MARCONIGRAMS: *KULCHUR'S* WIRELESS PROGRAMMING

T HE WIRELESS IN ALL ITS GUISES: a relay that connects *il duce* to a clique of followers, a connection whose entropy is converted into *parole in libertà,* and a technology that makes audible unheard frequencies by heterodyning them across the typed page or the invisible ether. Each is a function of the wireless, which struggles for predominance within a system of media. The modern poet occupies a privileged place as media go to war: he registers the new environment with its increased speed while his senses are routed through a relay that offers no time for reflection. Over time, pathologies and wireless technologies mirror one another in their ability to manage data streams so that Futurist F. T. Marinetti and Ezra Pound are prompted to register by their autism or state psychologists, respectively. The wireless connects poets to writing hands, typewriters, and secretaries who can take dictation, and the result is what I have termed wireless writing.

Pound's treasonous wireless broadcasts, the modifiers important for distinguishing his nonwartime transmissions of the last chapter from his more famous

Radio Rome broadcasts, provide further details on the affiliation of wireless technology and writing in the age of Marconi. Not only do the 105 transmissions Pound made over the radio from January 1941 to July 1943 display the above functions, but these transmissions depend on a particular detail that merits considerable attention: Pound's voice transmissions were stored on disks and played back in a carefully chosen sequence, so as to attain "their proper effect."[1] In the numerous examinations of Pound's broadcasts over Radio Rome, little space has been devoted to the technological media that made possible the storage and retransmission of Pound's voice.[2] This is odd, considering the importance Pound afforded the material storage of his voice on disks for garnering the attention of listeners. One of my chief tasks will be to examine the impact of voice storage on what Pound wrote for transmission, identifying its effects as the "technological correlative" of attention.[3] The partially connected media system sketched in the previous chapter is extended in these broadcasts to include a recording technology, the disk storage of voice, which by cutting allows for sequential manipulation. Switches among movement, speech, and action become the means by which listeners register Pound's wireless speech. For the wireless to avail itself of disk storage to trace new grooves in listeners' minds, two conditions must be met. First, the speed of the announcer's voice must reach a point at which his pulsations cause motion in the listener. Second, voice must become a source of data. Fiume's vibrating Marconi is fully realized only when stored voices can be transmitted because disk storage allows for the manipulation of vibrations, which is to say their speed could be increased, the effects noted, and motion created. In my reading, spacing and frequency modulation are the conditions for generating movement in bodies that interface with the wireless. What I hope to describe is an information system that reaches connectivity in Pound's broadcasts.

The texts I have chosen to analyze highlight the principal features of wireless programming. Chief among them is Pound's 1938 *Guide to Kulchur* for the connections he draws among attention, frequency modulation, and pedagogy, as well as the coordinates it provides for locating Pound's World War II transmissions within a self-generating wireless network.[4] To situate Pound's speeches within a more general context of wireless technology, however, I turn to Rudolf Arnheim's examination of an organic network of sound values from his 1936 text, *Radio*, especially his notion of a wireless voice economy. Taken together, Pound and Arnheim's musings on the wireless point to an aesthetic and ethic in which existence is directly related to one's capacity to be in motion. Or as Arnheim writes, "One only exists as long as one has a function, and if one's function is small, one's existence is small also!"[5] Second, there are Pound's FBI affidavits, recently collected in an anthology edited by his son, which I juxtapose with Mary de Rachewiltz's memoirs of her father's broadcasts published thirty years earlier.[6] Their principal interest lies in the description of the mate-rial conditions under which Pound recorded and transmitted his speeches, spe-cifically how indispensable secretaries and typewriters are for wireless writing. Finally, there are of course the speeches themselves, dictated, typed, and cor-rected by Pound and his secretary, ready for coupling with storage and trans-mission media. They are the site in which wireless commands are registered and performances commanded, of both Pound and his listeners.

When examining the speeches, I want to be especially attentive of the fol-lowing: What is the relationship between Pound and the network of which he is a component? How does Pound's poetry transmit energy and to what ends? Finally, what do voice storage and subsequent transmission mean for the poet who can now hear himself elsewhere, out of the place and time of the origi-nal recording, transmitted on a shortwave bandwidth? Friedrich Kittler speaks of dissolved feedback loops in his history of gramophonic effects on writing: "There is a feedback loop of logocentric notions, man hears himself speak, that is consciousness hears himself speak and sees himself write."[7] Appropriating Jacques Derrida's grammatology, he argues that with the development of voice-storage technology, media dissolve the formerly self-sustaining feedback loop of writing and speech. What takes the place of the logocentric feedback loop in Pound's transmissions on Radio Rome and where might we look for the effects of broken feedback loops? Consequently, other studies of Pound's speeches will appear primarily driven by theories of radio that privilege a certain logocentric notion of subject.[8] My reading overcomes these difficulties by emphasizing the spacing that underlies all meaningful hearing, the dictation and the registra-tion of data flows, and the connection between storage media and ears trained to pick up messages. Pound's speeches become an exhibit in a particular lineage of wireless technology that is not simply limited to their "spoken" qualities.

In the system of connected media that joins Pound's "Radio *Cantos*" to war commands, I suggested that fascism might be profitably viewed as a mode of wireless transmission that enlists bodies for warfare by short-circuiting the tra-ditional pathways of understanding. Unearthing the connection required an examination of the effects of heterodyned space on the typed page. With the ad-vent of technological sound storage, modulating frequencies to create attention on a much vaster scale becomes a reality. Frequency modulation and the altered time axis of acoustic events become one of the media conditions of fascism, which, I argue, holds true both for Pound's *Cantos* and for his radio broadcasts. Here I take my lead from Jeffrey Schnapp's examination of Futurist rhetoric and fascism, particularly his suggestion that a "set of interventionist theories and practices" later become important in the development of fascism.[9] I attempt to bring together systematically wireless programming and fascist codes of ac-tion. Both *kulchur* and fascism may be viewed as wireless systems that exploit spacing to inscribe commands in a flash, isolating the route to the ear in oracles and apocalyptic tones. The formulation is not without its difficulties, however, in accounting for what may be the culturally specific appearance of fascism. These I attempt to meet as the study closes, when I show how fascism models the dynamics of the wireless system by creating a feedback loop between sound manipulation and action. It exploits the wireless's penchant for winning over what Arnheim refers to as "hearts and heads."[10]

Arnheim's Radio Ethics

Spatial Voices

To begin, I want to return to Rudolph Arnheim's study of wireless listening in *Radio*. Written at the same time as Pound's *Guide to Kulchur*, it benefits from the author's intimate knowledge of Continental wireless communication networks that began to appear in the early 1930s. Arnheim was no engineer but he was a student of Gestalt who devoted himself to elaborating an understanding of art in terms of pattern recognition, all of which makes him an ideal foil for Pound's experiments in transposing data flows into different perceptual experiences. For Arnheim, the aesthetic configuration of wireless transmissions is based on an economy of the one voice, which creates listeners as it deprives them of the visual component of their perceptual sensorium. Three hundred pages of instructions for proper announcing over the radio amount to an injunction to keep uppermost in mind the economy the mechanized wireless voice institutes. "In wireless," Arnheim writes, "a single properly placed voice can fill the entire space and thereby express for the moment it is the only thing of consequence and apart from it nothing exists."[11] The question of whose voice and how many voices may be properly used is posed in the figure of the producer, whose duty it is to determine what effect is required to meet the demands of the occasion. The primary consideration will be occupying spatial depth:

> So it rests entirely with the producer whether the space is to be filled equally completely by a single voice or a plurality of voices. . . . In this way the weight and importance of an individual figure can be acoustically adjusted to any degree: a voice is not simply a voice, but according to its position in space, an impressive individuality or a mere nothing. Thus, through spatial conditions, the original acoustic equality of all human beings (as represented by their voices) gives place to a hierarchy determined by spiritual values.[12]

Arnheim's conflation of spiritual values through spatial manipulation of sound will not have gone unnoticed: it is one of the text's most illuminating characteristics, the ease with which the one voice comes to signify the voice of God, the

voice of the bodiless announcer, or, more menacingly by *Radio's* end, "the mur-
mur of a crowd and the shouting of voices . . . winding in a festive torchlight
procession through the dark streets of a big city,"[13] Arnheim runs a number
of relays between these considerations and a wireless broadcasting capable of
heightening the effect of the one voice: "Since you cannot give more than you
have, wireless will have to make good use of its few means."[14] Chief among them
will be relating the foregrounded voice to a particular hierarchy in which only
that which moves sounds. The competent wireless producer is one "who eman-
cipates the human voice entirely from its earthly appendage of material space
and lets it sound abstractly in nothing."[15] He must use care that "the listener
does not receive any unintended acoustic effects in these very simple wireless
performances . . . The number of aural phenomena should be deliberately limited in accordance with the law of economy we have already mentioned."[16] The
law of economy states simply that all media are "imperialistic" to the degree
each attempts "to give the whole situation" using only materials appropriate to
each.[17] As such, sound shifts between the one voice and nonessential acoustic
phenomena are to be avoided, precisely because the wireless medium does not
allow adequate processing of spatial depth. Judging the essential and the superfluous is not difficult if the only directed sound is the one voice exploiting
a wireless economy. When unintended acoustic events are a part of the transmission, voice and noise are understood all too readily as existing on the same
symbolic and spiritual levels. The reason concerns how meaning is processed
over the wireless. "Wireless achieves its effects not only by the succession of
sounds, but also by their superimposition," Arnheim argues. "Sounds which
are heard together are placed by the listener in relation to one another: the sensory coincidence suggests a relation of content." This is apparent when one considers that "individual things do not lie *beside* and separate from each other as
in visual space but *overlay* each other completely, even when, objectively, the
sounds come from different spatial directions" (emphasis in original).[18]

Note, however, the flip side of the superimposition of sounds. With the disappearance of the visual, "an acoustic bridge arises between all sounds: voices,
whether connected with a stage scene or not, are now of the same flesh as recita-
tions, discussions, song and music. What hitherto could exist only separately
now fits organically together: the human being in the corporeal world talks
with disembodied spirits, music meets speech on equal terms."[19] That it takes
one to know one in a media world, that only disembodied spirits speak with
each other, is a possibility Arnheim does not consider. On the one hand then,
we have the superimposition of sounds in a transmission that signifies by placement. On the other, there is an organic media network whose disembodying
effects on human beings allow for transmission regardless of time and context. Arnheim is suggesting that in an organic network of equal sound values,

information processing occurs by way of a "sensory coincidence" among sounds, which signify a relational rapport among them. What matters is what sounds, and by a neat trajectory between movement, sound, and the technical condition of blindness, what moves primarily is the voice.

To delimit this important point, I want to examine two further selections from *Radio*.

> Essential to what is happening at a certain moment is not so much the exis-
> tence of the inactive "being," but rather that which is changing, just happen-
> ing. . . . The best example of this is the human voice. It is silent when there is
> no action; when nothing is happening. If it speaks, it is to show that some-
> thing is going on. Activity, then, is of the essence of sound, and an event will
> be more easily accepted by the ear than a state of being. But this is the very
> aim of drama![20]

Information processing in Arnheim's wireless network occurs primarily by way of the human voice because the switches that run from silence to speaking bear primarily on whether an activity is being undertaken. The ear has difficulty in recognizing ontology or nonmovement, the "state of being" that has no need to be defined in relation to other moving bodies. Arnheim senses the outlines of information processing in the elegant functionality of sound data. "Modern mathematics has invented the notion of 'implicit definition.' That is to say, it defines the axioms which can no longer be traced back to higher notions by the functions which they fulfill in the highest theorems. Thus things are de-fined by their activity." Efficient processing of data occurs when the wireless, by superimposing sounds equally across a broadcast frequency, limits their func-tion solely to the relation they maintain with bodies in movement within the transmission. That Arnheim's "implicit definition" shares many features with Pound's use of the nonsyllogistic assertion in *Guide to Kulchur* is a point I will return to momentarily. Drama, Arnheim quickly adds, partakes of these same concerns: nothing about a man acting in a drama exists "that has not a function in a drama." But by installing blindness in the listener, the wireless amplifies the functionality of sound to a degree literally unheard of in drama. The dif-ficulty of the radio play (or the wireless transmission of more than one voice) is not in "ruling out the superfluous static," but rather "in including what is necessarily static in the action itself."[21]

The importance of Arnheim's insight cannot be overstated. In an aural world in which sound signifies movement, static too finds a mode of expression. Failing this, the success of the transmission is put at risk since superfluous static may have unintended consequences for data processing. The simplest source of static is the body. Although Arnheim glosses it as human character, he easily elides the term in favor of corporeal materialities. "It is this physical condition alone which first makes the actual utterances of the person comprehensible, and which must therefore of necessity be included in the aural drama." Of singular interest will be the speaker's tone. Including it in the "activity of actual speech," one will be able to hear the basis for the action in the action itself. Essentially, the tone of voice connects a listener with the specificity of the body. Then with a velocity that ought to give pause, Arnheim formulates a principle of art based on wireless exploitation of voice: "It [radio] sets 'existence' very clearly in relation to artistic function: one only exists as long as one has a function, and if one's function is small, one's existence is small also! This principle of art—which is also a principle of morality—is realised in the wireless much more radically than on the stage." One exists so long as one is speaking; anything less relegates the nonspeaker to the nonactive and hence the world of nonexistence. The reader waits another hundred pages to discover the full impact of the wireless on morality, but his or her patience does not go unrewarded. In a section appropriately titled "National Culture," Arnheim details the threat posed by weak administrations of wireless programming in both their "liberal" and "authoritative" forms. The former cultivate touchiness, are fearful of polemics, and in place of "courageous and effective statements" promulgate "abstract and guarded speech which certainly contains no shock but also gets nowhere." To illustrate his point, Arnheim calls on Mr. Edward Nutt, the chief editor of a newspaper in a story by G. K. Chesterton. "'He took a strip of proof instead, ran down it with a blue eye, and with a blue pencil, altered the word 'adultery' to 'impropriety,' and the word 'Jew' to the word 'alien,' rang a bell and sent it flying upstairs.'" To get somewhere in the wireless transmission seemingly requires a two-pronged strategy of synthesizing physiological characteristics into recognizable tonal patterns and calling aliens by their proper name.[22]

Now, a reference to noted British nationalist anti-Semite G. K. Chesterton might simply be an interesting coincidence were it not for the ease with which the wireless demands that friend and foe, proper and improper, true and false be clearly demarcated.[23] Summing up the wireless direction two pages later, Arnheim writes:

> Wireless can be directed quite consciously from a definite point of view. The principle of choice is: what is for us and what is not? The question of quality, of cultural standard, must occasionally take second place. Diversity here stands for characterlessness, bias as the natural presupposition of all cultural activity. What does not fit into the policy is either passed over or represented in a negative sense.[24]

Arnheim remains supremely confident that the wireless meekly responds to humans that transmit over it. But if his study of tonal physiologies prompted by the wireless proves anything, it is that the wireless viewpoint is ignored at one's peril.

Sound as Medium

This perspective requires no introduction: it is the same one I focused on when describing how technical media work together to devise strategies for capturing the real. In chapter 3, I showed how the wireless was one of the technological conditions for the production of *parole in libertà*. With the advent of voice transmissions, voice recognition became the meager hope for averting the entropy of the wireless registration. The stark "for/against" that relegates cultural standards to second place functions similarly since it too measures the entropy of the circuit running between speaker and wireless while simultaneously introducing information of a specific kind. Arnheim identifies the information early in his study: "We must not forget, however, especially when dealing with art, that mere sound has a more direct and powerful effect than the word. The meaning of the word and the significance of the noise are both transmitted through sound, and have only indirect effects."[25] Sound is the medium through which the meaning (of the word) and the significance (of noise) are transmitted. Sound is the "sensuous unity" that treats the word and noise equally.[26] Yet, Arnheim's difficulty in conceiving of noise without meaning and his failure to specify, consistently, the origin of sound should not prevent us from drawing the appropriate conclusions.[27] In the wireless interface, the amplified sounds of the machine (voices) are connected to the human ear and then lost energy (entropy, uncertainty, call it whatever occurs when machines and users are joined) is converted into information regarding their frequency and the possibility of acoustically adjusting the weight of an individual figure.[28] This represents an advance on Marinetti's manifestos and Pound's "Radio *Cantos*," for now the wireless regenerates itself by giving importance (Arnheim would add the qualifier "spiritual") to transmitted voices. Energy is reintroduced to generate greater voice recognition capacity. Arnheim fashions an aesthetic strategy based on the possibility of modulating frequencies among voices.

What is the connection between these self-organizing processes and wireless information processing? Does greater voice recognition *by definition* increase the capacity for identifying friend and foe, patriot and enemy? Clearly the friend/foe distinction existed before the emergence of the wireless. Manuel de Landa describes how Napoleon exploited the vast reservoir of human resources, which the French Revolution had equipped with a heightened capacity for determining friend and foe. These resources powered the first motorized army in history.[29] The wireless also successfully installs a friend/foe distinction but on a vaster scale by creating conditions under which voice recognition prospers. Since voice recognition proceeds, via Arnheim, in terms of who moves and who, by not moving, is exiled to nonexistence, activity functions as a switch that is capable of being turned "on" or "off" in response to other switches being "on" or "off."[30] In wireless transmissions, where sound equals movement, those that sound are the same as those that move.

Thus, the wireless may be represented by the flow of information that oc curs when the switch between activity and nonactivity is turned on in response to another switch, in this case sound and nonsound, also being on. In short, a wireless system exploits the friend/foe difference and successfully switches between the production of sound and activity. The wireless network allows this flow of information to organize itself, and a contrast is enacted as tonal opposi tion between voices, which is later directed to ears that can apprehend them as a "primitive acoustical contrast."[31] These primitive acoustical contrasts are the first step in creating the binaries of for and against, which will coincidentally also provide the transmission with its addresses. We can specify the address: inasmuch as a wireless voice transmission operates via tonal opposition, its destination is primarily to the "listener's heart and head, because they alone answer to the particular nature of the broadcast."[32] One begins to sense a lineage of wireless technology running from Sergi's experiments on spacing as a condi tion for meaningful hearing to the development of the superheterodyned signal in World War I to greater voice recognition as outlined by Arnheim. Each wire less moment successfully transforms bodies into listening subjects and gener ates knowledge about the body and its speaking apparatus. In Arnheim's case, we can express the importance this way: acoustical adjustment of voice and the leveling of sound values caused by superimposition function as commands that make data processing possible, which represents a leap in the technical possibilities of the wireless. To achieve proper connectivity, however, storage operation is needed. It is to this I now turn.

Gramophonic Wireless Transmissions

Information processing in the wireless occurs when ears register tonal opposi tion of voices as signs that identify the origin of the voice in the body.[33] A fur ther step in the development of the wireless network occurs when voices can be stored. Arnheim describes how the wireless might best utilize the advantages of technological voice storage:

> The ease with which wireless can present occurrences at various places and times as a unity and in spatial juxtaposition is especially suggestive if it deals not only with imaginary themes, but also with genuine ones taken from re ality. If, for instance, sound-shots of various episodes from the life of a politi cian existed, they could be put together in a sound-picture and so a whole life could be concentrated in a single hour. . . . Wireless directly juxtaposes what is farthest removed in space, time and thought with amazing vividness.[34]

The move from imaginary themes to "genuine ones taken from reality" implies storage devices that Arnheim's organic network will exploit to meet fully its wireless possibilities. For sound shots (the shooting of cameras, microphones, and guns is overdetermined throughout *Radio*) to reach their intended receiver, they must first have been stored as data in a network capable of coupling them with a transmission medium. By the time of *Radio's* writing, the history of voice storage was already fifty years old. Edison's experiments at Menlo Park had led to the primitive gramophone, a device that could inscribe the frequencies of voice onto disks to be played back at another time, respecting only the condition of the inscribed surface. Indeed, Marconi often had music piped in wirelessly onboard the *Elettra* as he criss-crossed the globe, transmitted to him from the various shore stations that made up the Marconi Company's wireless network. Future gramophonic improvements included electromagnetic cutting amplifiers for recording and an electromagnetic pickup for replaying. Eventually, as entertainment networks grew up out of the ashes of World War I receiving equipment, the German company Siemens offered "the recording studios of the media conglomerates with equally electric ribbon microphones, as a result of which grooves were finally able to store frequencies ranging from 100 bass hertz to 5 kilohertz overtones, thus rising to the level of medium-wave transmitters."[35] Thus, preparing gramophonic recordings for later wireless broadcast in the period Arnheim describes required that overtones come together around certain low frequencies.

As long as the gramophone was capable only of storing the limited range of the human voice, however, its use for the wireless was circumscribed. Increasing the sound fidelity of voice storage created greater connectivity with the wireless, though the development of high fidelity was still a decade away. This explains Arnheim's statement that "for years, composers for gramophone, sound-film and wireless have been aiming at a type of instrumentation that limits the volume and fullness of sound to an average level."[36] Mixing imaginary and "genuine" sounds assumes then a convergence of frequencies in voice transmissions and technological storage.[37] The effect will be dramatic, making us "independent of the time and the place of production" because "original records of real sounds as well as of historical events can be mingled"[38] The process occurs when sounds can be cut and mixed, which results in some interesting temporal effects, what in wireless parlance is known as time-division scramble or TDS. With the development of voice scrambling systems in the 1920s and 1930s and the greater use of the sound strip for radio, the stream of speech could be chopped up into split-second portions and then randomly shuffled. On the other end was a descrambler to put the sounds back into their proper order. Speech was no longer simply transformable in its frequency but could now be enciphered by altering time. TDS uses systems that encipher "by changing the temporal relationships of speech's continuous flow." They must preserve it "momentarily to permit the transposition. Usually they have used magnetic tape."[39] The development is significant, for the wireless no longer depends on the inspiration that occurs at the moment of transmission but can now superimpose "present" acoustic events on those of the past, inspired or not. Arnheim introduces the term "broadcasting" to identify the moment wireless and recording technology are coupled, which is generally consistent with media histories of the period.[40] The development of the sound strip followed shortly thereafter, allowing for greater manipulation of sound data through a process of acoustic montage. "It would be a step of great importance for the development of the art of radio drama," Arnheim muses, "if every radio play which used space and montage as a means of expression were not 'performed' in the studio as if on stage, but were recorded piecemeal on film-strips like a sound-film, and the individual strips cut properly afterwards and mounted as a sound-film."[41] Sound strips offer the producer "the possibility of exactly determining the 'cutting,' that is, of specifying the beginning and the end of every piece of montage exactly to a second."[42] Their introduction will make possible the appropriate fading in and out necessary to the wireless, and will help separate one broadcast program from another. Most important, by offering the engineer the possibility of precise sound cuts and the capacity to recombine them to create effects, they create another source of acoustic data.

My brief encapsulation of the wireless-storage hookup is in accord with Kittler's judgment that greater connectivity between sound storage and wireless transmission dramatically altered the media environment.[43] What is distinctive about transmitting recorded sounds wirelessly and how does another medium, poetry, react when faced with a fusion of this sort? Consider first the effects of sound storage alone on temporality. Sound inscription made possible time-axis manipulation, what Arnheim mistakenly calls freedom from time, as one could now quite easily augment the speed at which the gramophone played back its inscribed surface; sound strips of the sort the wireless engineer used functioned similarly. In each, playback at different speeds required only a needle and an amplifier since nothing prevented one from retracing the inscribed surface at a faster speed: inscribed vibrations that begin a record could be manipulated so as to be heard at the end. In addition, the temporal unit in which the acoustic event took place could be altered by increasing playback speed. All this will appear quite obvious to anyone still owning a 33 rpm record player. Less obvious is the process in which gramophonic inscriptions become programmable acoustic events, equal in stature to those "live," since they are superimposed at the moment of transmission. Only when acoustic events are recorded in a data form compatible with the wireless, that is, inscribed with an "average level" of volume and sound, is superimposition possible. This is because the wireless hookup before high-fidelity functions one-dimensionally in terms of distance: "All those different spatial characteristics of sounding bodies in the transmitting-room are reduced, in their effect on the blind listener, to his hearing, along one extension in depth, sounds coming from various distances." Therefore, the wireless combines acoustic data, exploiting the possibilities the gramophone provides for time-axis manipulation by making the inscribed voices its own. Past (inscriptions) and present (transmissions) will come together to weaken a strict conception of linear time since "the process of exchanging words is only comprehensible when they or their bearers are imagined as existing at the same time *side by side*" (emphasis in original). In order for processing to occur, however, the visual must disappear. Common to the wireless and the gramophone is the sensory quality of sound, suggesting an origin in their respectively installed blindnesses.[44]

Every blind listener requires a dog, which the wireless happily provides. The reporter, "the blind man's dog accompanying the helpless listener," effectively transmits when he can improvise "coherent and vivid information" at the microphone, which means "resigning 'the word' in favor of the sound of the thing at the right moment."[45] When done properly, distance may be conveyed and attention created via artificial cutting: "It [the relay] conveys distant happenings to the listener by the most direct method conceivable to-day, that is to say, it 'artificially cuts out slices of reality' by this isolation making them the object of special attention, sharpening acoustic powers of observation and drawing the listener's attention to the expression and content of much that he ordinarily passes by with deaf ears."[46] The wireless-gramophone hookup isolates sound data, be it the reporter's voice or the sound strip, generating attention and promoting more acute hearing.

Yet we may express the relationship differentiated voices maintain with each other more precisely. The media linkup establishes differentiality through a topology of the expressive sound, which orders both the transmission and its reception. Voices become acoustic data that take on meaning only when heard together with contiguous sounds. Thus, the wireless broadcast practices a technique of sparseness, a logic of the signifier that releases a maximum amount of energy through a minimum number of signs/sounds because it will be "impossible to superimpose a multiplicity of texts, or they will become incomprehensible."[47] Note too that it is the differentiality of the sound cut that precedes meaning, made possible by the exclusion of the visual. The blinding power of the wireless isolates acoustic signifiers from their optical signifieds, grounding them against a background of noise. Their materiality resides in the cut that identifies them as difference, what David Wellbery in a not so different context helpfully calls "a labor of differential inscription that is both prior and irreducible to meaning."[48]

We can now trace the outlines of a wireless network: gramophonic inscription and the sound strip provide technological sound storage, primitive acoustic contrasts address transmissions to ears programmed to pick them up, and data processing operates in the contingent space of presence/absence that separates and unites the succession of sounds. The wireless network alters the structure of other transmission media, which in turn register the new media ecology. Yet one wonders what happens to another transmission medium when the wireless utilizes the gramophone to transmit the past into the present, devaluing words in favor of their real aspects, having recourse to a typology of voice and its contrastive tone? I examine this knotty question in the following pages by turning to Pound's *Guide to Kulchur* for a possible answer.

Pound's Kulchural Studies

The Algorithm of Understanding

In the altered media ecology, poetry avails itself of a wirelessly inflected materiality that conditions its possibilities for transcription in the hearts and heads of his readers. Pound's *Guide to Kulchur* is exemplary. Written in 1938, it is considered an essential expression of Pound's aesthetics, be they fascist or simply ideogrammic. In the chapter "ZWECK or the AIM," Pound lays out the parameters by which man will be "gittin' Kulchur." The process begins with the transformation of individuals into registering machines: "The ideogrammic method consists of presenting one facet and then another until at some point one gets off the dead and desensitized surface of the reader's mind, onto a part that will register."[49] Pound correlates a method of presentation, "presenting one facet and then another," and the same method's capacity for finding the registering surface in the reader's brain. He will vary in his choice of descriptors for a writing that inscribes itself onto the sensitized surface of the brain: there is the ideogrammic method mentioned here; the "enthusiasm" he speaks of when attempting to explain Plato's success with his disciples; and the escape from "a word or a set of words loaded up with dead associations."[50] However he identifies these forms of writing, each is characterized by its relation to memory.

When Pound first describes the ideogrammic method some thirty pages earlier as "a type of perception, a kind of transmission of knowledge attainable," he opposes it to another transmission, that of writing down and memorizing. "May I suggest," Pound writes, "that I have a certain real knowledge . . . and that this differs from the knowledge you or I wd. have if I went into the room back of the next one, copied a list of names and maxims from Fiorentino's *History of Philosophy* and committed the names, maxims, and possibly dates to my memory."[51] The perception that undergirds an ideogrammic method depends on an information technology different from that of the copying hand, which, when connected to memory, allows the body to commit to written particulars. Instead, the ideogrammic method "remains effortlessly as a residuum, as part of a total disposition." The demand for a different kind of transmission technology grows acute, "coming even closer to things committed verbally to our memory."[52] That is, as Pound attempts to commit voice to memory (his own or another, it matters not because only vibrations are recorded), and not simply the names and maxims of a philosophic tome, the limits of the previous media linkup become clear: a copying hand fails to inscribe into memory verbally committed details.[53]

Unfortunately, Pound fails to assemble the mechanism by which the ideogrammic method inscribes sound events into memory. He soon writes, "I haven't lost my thread," by which point he has, but not before gesturing to the primary component in the linkup: "I may, even yet, be driven to a chronological catalogue of greek ideas, roman ideas, mediaeval ideas in the occident. There is a perfectly good LIST of those ideas thirty feet from where I sit typing."[54] Pound *types* his disavowal of chronological catalogs, which are the product of an outdated transmission from memory to hand. In its capacity to set off, typographically, "LIST" from "greek," "roman," "mediaeval," and "occident," the typewriter separates Pound's body from the paper on which he writes, for he no longer need commit catalogs to memory through the hand/memory linkup. A still uncertain process that runs through the non-handwritten frees him from bodily data storage.[55] Thus, two media links store data, though only the one utilizing the typewriter has the capacity to record verbal inflections.

Central to Pound's understanding of the new transmission is the difference between knowledge and understanding: "Knowledge is or may be necessary to understanding, but it weighs as nothing against understanding, and there is not the least use or need of retaining it in the form of dead catalogues once you understand process."[56] Understanding is joined metonymically to weightlessness; once "the process," Pound's code for the ideogrammic method, is understood, "it is quite likely that the knowledge will stay by a man, weightless, held without effort." This occurs when memory is no longer connected to a hand that writes cursively but to a typewriter. The typewriter frees the poet from the laborious task of storing knowledge by separating the flow of writing from its acoustical counterpart, which the handwritten, copied text so easily conjoined.[57] Memory, or better, the connection running between hand and consciousness, will be needed less.[58]

One also notes that the weight Pound's catalogs carry, especially on the eyes, allows the bodily requirements of data storage to be calculated. Recalling a day passed at the British Library, Pound is able to quantify "the eye-strain and the number of pages per day that a man could read, with deduction for say at least 5% of one man's time for reflection." Deciding against it, he opts rather for a "knowing that is in people, 'in the air,'" for an archaeology not retrospective but "immediate."[59] Reading, copying, and memorization are part of an earlier media ecology that Pound abbreviates as knowledge, which required a prerequisite eyestrain in order to function: knowledge derives from a pain that brings forth memory.[60] Pound's diatribe against cataloged knowledge presupposes the eyestrain that a typewriter relieves, though the typewritten inscription too is not without its pain-producing mechanism. Under new media conditions, the second hookup operates acoustically "in the air" and "in people." Perhaps the reader will recall the weightless rhetoric I associated with Marconigrafia in chapter 1, especially the bypassing of matter in favor of immediacy and mental telepathy that Marconi's device supposedly set in motion. Pound's category of understanding operates wirelessly to the degree it disconnects the visual hand/memory circuit in favor of an acoustic (and blind) connection that is immediate and whose bodily storage requirements are slight. This is the other side of the wireless coin: heavy matter, be it bodies or cables, can be bypassed by a technology that more efficiently and effortlessly inscribes its commands onto a registering surface.

Put another way, the typewriter separates memory from the handwritten. As a result, a poet need no longer move to where the books are but merely type what his or her antenna picks up out of the air. In such a scheme, understanding becomes an algorithm of a data flow sent across a wireless circuit and managed by a typewriter. This is not to deny that random noise, the background of all media, does not travel across the circuit, but only that understanding is always a selection from a more complex whole, be it the Lacanian real or Luhmann's environment. Pound expresses the idea succinctly some pages later: "It is not what a man says, but the part of it which his auditor considers important, that measures the quantity of his communication."[61] Indeed, *Guide to Kulchur* is dedicated to the idea that in 1938 poetry could, by getting off "the dead and desensitized surface" of the mind, reach that part of the brain capable of registration. In the remaining pages, Pound searches for the most efficacious tools for inscription, while detailing the requirements for the successful programming of his readers/auditors.

Textual Mechanics

The search for programming tools begins with the assumption that speech and action measure energy. In a chapter entitled "TOTALITARIAN," Pound writes: "Whatever the platonists or other mystics have felt, they have been possessed sporadically and spasmodically of energies measurable in speech and in action, long before modern physicians were measuring the electric waves of the brains of pathological subjects. . . . There is no doubt that Platonists, all platonists every Platonist disturb or disturbs people of cautious and orderly mind."[62] Pound is describing models of energy transfer. In the first, energy possesses mystics "spasmodically," that is, they gear into the body, where speech and action measure the degree to which they circulate without hindrance. In the second, the electrical waves of "subjects" will demonstrate how effectively pathologies increase the production of electricity in the brain. In both cases, brains prompted by Platonism and pathology electrically signal their presence within a system in which energy is transmitted. The mystical or pathological subject, Pound seems to argue, is the result of a transmission of energy that "concretizes," to borrow Lacan's formulation, "its formation."[63]

Pound, however, fails to account for the actual production of energy in the two models, focusing instead on the historical consequences of his insight. The history of "kulchur" will be reduced to the electrical pulses that manifest themselves as "ideas going into action."[64] Inasmuch as fascism represents ideas "going into action," it literally electrified its subjects. In order to identify the process whereby energy transfers augment electrical waves (and so condition subject formation), I will assume, not unproblematically, that *Guide to Kulchur* functions as a textual machine capable of transmitting and producing energy. There are good reasons for this assumption. First, Pound understood his writing, both the prose and the poetry, as a code for action; he even devoted a book to machine art that could produce energy.[65] Second, one recalls Lacan's reminder that if a rabbit is to be pulled out of a hat, "it's because you've put it there in the first place."[66] Subject formation proceeds because a machine has been set in motion: it is complete when the machine stops where it has been programmed to stop, whether the machine be a wireless signaling device, a shortwave radio, or a sequence of typographic signs. Pound's pedagogical writings, of which *Guide to Kulchur* is one, are especially helpful in mapping the programming features of poetry that flashes its way across the page and onto a bodily surface.

Consider again the descriptors Pound uses to describe the effects of energies manifested in mystics: first, they occur sporadically, which suggests that the energies are transmitted in intervals. So, the kind of machine at work here is capable of regulating energy flows. Second, note that possession is localized on the body and manifests itself in spasms that result from the same flow, which indicates that switching from one channel to another is not only possible but necessary. Finally, there is Pound's choice of the plural "energies," confirming that more than one channel is in use. With these pieces of information, we can now specify the kind of machine in operation as a Boolean motor:

The Boolean motor is embodied in most systems in which energy of any sort is transmitted through a network of channels, with devices that can turn the energy on or off, and switch it from one channel to another. . . . The energy can be a flowing gas or liquid, as in modern fluid control systems. It can be light beams. It can be mechanical energy transmitted by wheels, levers, pulleys, and other devices. It can even be sound waves or odors.[67]

One of the most significant incarnations of the Boolean motor will be in the control of the flow of electricity inside computers. It was Claude Shannon who demonstrated how relay and switching circuits could be expressed by equations using Boolean algebra: "In these equations, True and False correspond to the open and closed states of a circuit."[68] By substituting the binary connectives, "and" and "or," electrical circuits could be designed using typographical resources to produce energy. Is the same possible for determining the kinds of circuits producing energy flows in Pound's pathological subjects and mystics?[69] If one were able to identify the Boolean motors responsible for creating energy in *Guide to Kulchur,* one could show how nonproductive operators are broken down and combined in such a way as to produce motion and energy.[70] Not surprisingly, we have been preceded by Lacan.

In his seminal essay "Sign, Symbol, Imaginary," Lacan deploys a Boolean motor in accounting for how sense is brought to language:

[I]n order for the message to be a message, it is necessary not only that there be a succession of signs but that this succession of signs has a direction, an orientation. In order to function according to syntax, it is necessary for a machine to move in one sense or another. And when I say machine you can see that I am not talking about some little thing which we usually call a machine. When I write on a piece of paper the transformations of I's and O's, this production always has a direction.[71]

Lacan's intent in isolating the symbolic function of sense against the imaginary order need not concern us here so much as his emphasis on the movement of a machine in the direction of one sense or another. The difficulty of the passage is lessened if we read sense not simply as meaning (though clearly Lacan means that too), but more properly as corporeal media (visual, acoustic, tactile). The senses are the interface between the binary operators (I and O) and what the program is intended to do. Thus, the succession of signs that form a message are directed both to making sense and reaching the senses, which explains why mistakes in syntax alone are less useful than mistakes in programming: "Mistakes in programming will engender falsehood" because programming depends on the play of true and false.[72] Therefore, programming results from a motor capable of noting the direction in which a succession of signs moves by transforming them into I's and O's. A sequence of switches that open and close according to the direction that true and false indicate follows in due course, suggesting that the flow of truth which results from, for example, deduction can be variously reassembled to create energy (and not simply transmit truth) when its components are recombined in a way that produces motion.[73] Perhaps motion is too ambiguous a term for what these motors produce, so we might take it here to signify a kind of flexibility productive of action or even a formation of data.

Revelation is the name Pound gives to the process that electrifies bodies into subjects. "What we can assert," Pound writes, "is that Plato periodically caused enthusiasm among his disciples. And the Platonists after him have caused man after man to be suddenly conscious of the reality of the *nous*."[74] The nature of revealed reality occurs to the mind of the disciple and results in that most prosaic of effects, enthusiasm. Given what we know about oracles, revelation and vibrations, it will proceed "suddenly" via writing and media.

> At last a reviewer in a popular paper (or at least one with immense circulation) has had the decency to admit that I occasionally cause the reader "suddenly to see" or that I snap out a remark . . . "that reveals the whole subject from a new angle."
>
> That being the point of the writing. That being the reason for presenting one facet and then another—I mean to say the purpose of the writing is to reveal the subject.[75]

When the media ecology was dominated exclusively by writing's storage and transmission capacity, writing occupied the position of Dante's God transmitting truth to man.[76] With the advent of separate data flows and wireless circuits, writing will require other operators to effect revelation. In 1938 the function of writing resides in its capacity to make transmittable "one facet and then another" of a subject. In each environment, however, only the transmittable is revealed, which is to say with Lacan that the fundamental relation of man to the symbolic is the same one that creates the symbolic order itself, "the relation of non-being to being."[77] So the task becomes identifying the operators of writing in *Guide to Kulchur* that are available as typographical options to program those seeking *kulchur*.

. . . And Typographical Operators

Whereas the crucial operator in the "Radio *Cantos*" was the quotation mark that made possible switching between voices and centuries, Pound's pedagogical writings (*ABC of Reading, Guide to Kulchur, Polite Essays*) utilize typographical space to juxtapose and disconnect statements and paragraphs so that what reverberates is not merely the sense of statements but the tone revealing an on-going disclosure. Pound outlines this method repeatedly, but nowhere is it more explicitly linked to operators than in the chapter "Sophists":

> The syllogism was counted as "merely a grammatical form." It didn't pull any weight. Was merely useful for hypothesis and dissociation. Didn't PROVE anything. In that position lies the intellectual greatness of the school of the Porch. That the syllogism was not apodictic but anapodictic. After all the greek blather. Here they got their teeth into something. Their four categories were: 1 the substratum, 2 the general quality, 3 the determined modification, and 4 the relatively determined modification. Each presupposing the one named before it.
> I don't know that they need any plus marks here.[78]

Expressed in a syllogism: if a syllogism is incapable of demonstrating anything (anapodictic), then what remains are statements that can be combined and/or associated in such a way as is useful for pulling weight. In the typographical space separating statements that do not combine to create truth on paper, information circulates as "the presence/absence of absence/presence" independently of any subjectivity: Pound's reader is indistinguishable from a simple machine programmed to perform calculations.[79] One need only symbolize the necessary operations (perhaps as plus marks as Pound himself notes) and a condition would be met for a man "gittin' Kulchur." Central to the circulation of information is therefore a machine whose general data processing is implemented in the space between statements.[80] It is a short step from a program that pulls weight to a condensed poetry whose task is to inscribe *kulchur's* program onto a surface that will register. This is because poetry, unlike prose, processes data much more efficiently: "There is MORE in and on two pages of poetry than in or on ten pages of any prose."[81]

Apparent throughout is the degree to which spacing articulates the becoming-space of time, what Derrida calls "the unperceived, the nonpresent" in constituting the subject of *kulchur*.[82] In terms that will help determine what is at stake in linking data processing and typographical operators, poetry is the result of a feedback loop between spaces (or symbols if we choose to write out the plus signs) and letters, which together bring about *kulchur's* self-organization. There are a number of good reasons for seeing *kulchur* as an autopoetic system, which will become clearer as I proceed, but we can anticipate briefly the process whereby man is generated as *kulchur's* output. This occurs only after the operations of *kulchur* are repeated; only after man has been programmed to "escape a word or set of words loaded up with dead association."[83] The system is autopoetic when its output, "kulchured" man, becomes its input in a recursive loop. Indeed, we might say that the interplay between space and letter is itself a medium that serves as an interface between man and a self-generating social system, *kulchur,* and permits their coupling. The spacing poetry utilizes when in competition with other media makes it especially adept at encoding the difference between information and the individual utterance, a primary condition for acquiring and processing information.[84]

Thus Pound writes that "man gittin' Kulchur had better try poetry first."[85] More so than prose, poetry exploits the possibilities of spacing and with it the difference between information and utterance. It does so by repeatedly addressing itself to readers who are new to the game: "The 'new' angle being new to the reader who cannot always be the same reader. The newness of the angle being relative and the writer's aim, at least this writer's aim being revelation, a just revelation irrespective of newness or oldness."[86] There appears to be some genetic component of poetry that allows for the processing of information regardless of time (whether it is new or old) and with a speed sufficient to escape a dead association. Along the same lines, Pound will later call poetry "a sort of inspired mathematics, which gives us equations, not for abstract figures, triangles, spheres, and the like but equations for human emotions."[87] Central to man "gittin' Kulchur" is the notion that one acquiring and processing information recognizes to some degree the difference between what cannot be captured symbolically, Pound's chaos of the last chapter, for example, and *kulchur,* which represents a selection of the former. "Just revelation" operates in the twilight zone between an uncapturable real and *kulchur'*s algorithm. It requires a deficit in the reader's brain (the "dead and desensitized surface" on which to inscribe itself) and a binary machine that exploits spacing in order to locate a surface receptive to its commands. I add in passing that these operators function strategically since they are legible only to a machine whose programming allows tampering with the relationship between itself and its pattern recognition protocols.

Two caveats should be immediately noted. First, I do not want to suggest that all binary schemes are limited to revelation as it is clear that they are built into the "horizontal structure of all meaningful experience: to continue on or break off."[88] Revelation indeed shares many of the same features of other communicative events; it presupposes meaning that enables the interpenetration of systems (*kulchur* and man "gittin'" it) and its articulation. Where Pound's revelatory program succeeds is in the efficacy with which it readily connects a behavior on the part of man with the possibility favored by *kulchur,* which is to set motion in motion. Revelation, in other words, uses speed to shrink time and so fails to reward reflection. Second, the subject of *kulchur* may be designated only in relation to the realizations that Pound's binary machine holds open; these are the result of a series of operators effecting "right" movement and right interpretation. Risking a tautology, "kulchured" man is the result of its inter-penetration with *kulchur,* since it is interpenetration that selects the structures "that enable the reproduction of interpenetrating systems."[89] *Kulchur* is nothing without its going into effect.[90] It does so by recombining the set of signifiers that is man in order for its own reproduction. In his bitter opening note to the 1970 edition, Pound sums up *kulchur*'s consequences on meaning and man:

> *Guide to Kulchur:* a mousing round for a word, for a shape, for an order, for a meaning, and last of all for a philosophy. The turn came with Bunting's line:
>
> "Man is not an end-product,
> Maggot asserts."[91]

Essentially, the search for the lightning-fast shape or order that generates *kulchur* can help determine where any errors in programming lie: a man "gittin' Kulchur" becomes the means by which the proof of statements may be dem-onstrated. But the hunt for an order and a meaning need not assume man as meaning's end product.

A Test of Broadcast Systems

The need for such proofs grows more acute the less writing functions as the sole storage medium. In a print-based media ecology broken apart by technological voice storage and transmission, truth is only transmitted with difficulty by the printed page. "Truth is not untrue'd by reason of our failing to fix it on paper," Pound obsessively writes near the end of the *Guide to Kulchur.* "Certain objects are communicable to a man or a woman only 'with proper lighting,' they are perceptible in our own minds with proper 'lighting,' fitfully and by instants," he adds.[92] For Pound truth arrives only when accompanied by the lighting poetic form provides in the instantaneous flash that paper fails to store: hence truth is not stored but revealed. When discussing Hardy's poetic genius, Pound again names the transmission that operates in a flash, lighting memory and transmit-ting truth. "No man can read Hardy's poems collected but that his own life, and forgotten moments of it, will come back to him, a flash here and an hour there. Have you a better test for true poetry?"[93]

Note what Pound's remarks assume. Traditionally, poetry had served to reach into the brains of its listeners via their ears in order to improve the storage ca-pacity of memory. Under oral conditions, it worked as a substitute for memory. "Oral verse was the instrument of cultural indoctrination, the ultimate pur-pose of which was the preservation of group identity. It was selected for this role because, in the absence of the written record, its rhythms and formulas provided the sole mechanism of recall and re-use."[94] The test (and task) Pound gives poetry varies according to the respective media ecologies in operation. With the advent of technological storage media, Pound seems to say, less storage capacity will be taken up with catalogs of "the names of protagonists, or authors of books" since there are now machines to record voices and cabled and wireless possibilities to transmit them.[95]

The question really concerns the impact a novel medium has on the operative capacity of a network already in use. Derrida poses the problem similarly in an extended footnote to part I, chapter 3 of *Of Grammatology*. Quoting approvingly from Leroi-Gouhhran, he argues that the loss of printing will not be regrettable "since printing will conserve the curiously archaic forms of thought that men will have used during the period of alphabetic graphism; as to the new forms, they will be to the old ones as steel to flint, not only a sharper but a more flexible instrument."[96] The nod to possible friction between steel and flint is in accord with one of this study's fundamental tenets. As wireless media expand their reach, friction is produced with the previous information system that, in the case of Pound, is primarily writing-based. The cultural history of the interwar period can be read then as a competition between devices of transmission and storage, isolating or juxtaposing moments of extreme friction.[97]

What are the consequences of such a view for poetry, that is, under heightened media competition, what is poetry's function? No longer simply a mnemonic device or the effect of the circuit running between the Muse and writing, poetry becomes, under wireless competition, "a code for action."[98] It sets in motion a process whereby "one is led or edged over into considering them [words] with greater attention."[99] Our foray into Arnheim's wireless aesthetics can help specify the nature of the code that creates action by way of generating attention. Consider again how the wireless-gramophone hookup isolated acoustic data so as to draw the listener's attention to the sparse, expressive sound. It took on meaning only when heard across an acoustic bridge. Under conditions of friction with wireless media, Pound's *Guide to Kulchur* utilizes the only means available to it, typographical spacing, what Pound in another context calls "blank words" to juxtapose statements, thereby disclosing tones and making operational an information processing that no longer works hermeneutically on signifieds.[100] To the extent *kulchur* exploits spacing to inscribe commands in a flash, it operates wirelessly. Poetry becomes the wirelessly transmitted voice whose very isolation contributes to its capacity for issuing commands. What happens, however, when a poet already typing wireless messages is called by the wireless to speak over its airways? Naturally, further details on the partially connected media system in operation become apparent: it begins to resemble what we might call a wireless command structure.

Pound and Voice Storage

News from Nowhere

The details of Pound's broadcasting career are well known. From January 1941 until September 1943, when Marshall Badoglio became head of the Italian government, Pound spoke twice a week over Radio Rome. The texts he wrote at his home in Rapallo and "on occasion in Rome where he traveled to record on discs a batch of 10 to 20 speeches."[101] The speeches were typed and included handwritten changes, the indicated target audience (Great Britain or the United States when deemed necessary), and the numbering of the broadcast sequence. Each speech began with the famous formula absolving Radio Rome of any coercive influence over the contents of the broadcast and then proceeded, in a sort of hit parade of villainy and anti-Semitism, to attack those he considered most responsible for the war. These shortwave broadcasts, often heard with difficulty and at strange hours, were monitored by the Foreign Broadcast Intelligence Service of the Federal Communications Commission, which in conjunction with a Grand Jury, indicted Pound for treason in July 1943.[102] The U.S. Counter Intelligence Corps headquarters in Genoa issued an order for his arrest soon after the broadcasts ceased. Pound was arrested in May 1945.

In a less cursory examination, we might note that Pound's first broadcast on Radio Rome actually occurred six years earlier in January 1935, when he spoke "on invitation" from Italian Foreign Minister Ciano for the shortwave program *The American Hour,* which included conversations with such luminaries as Enrico Fermi and Guglielmo Marconi. The title was "Conversations with America on How the *Duce* Will Resolve the Problem of Distribution."[103] Interestingly, de Rachewiltz has Pound's history with the wireless beginning even earlier, in 1931, when radio "caught his attention." He heard his opera *Villon* "in the electrician's kitchen" of Rapallo and could do no less than to exploit this new means for "the diffusion of knowledge." She also observes that after Pound received Natalie Barney's gift of a radio, made famous in his letter to Duncan of March 31, 1940, he became "an avid listener, and learned to differentiate voices. . . . The urge to respond overcame him, although he was aware of his disadvantages."[104]

The reader will certainly recognize in this equipmental compulsion a link to our assembled cast: an adolescent Marconi, D'Annunzio at Fiume, and Marinetti practicing *parole in libertà*. In addition, broadcasting gave Pound the opportunity to practice ventriloquy. "I like Morelli's reading of my stuff," Pound writes approvingly. "The anonymous stuff is in some ways better than the personal / When anonymous I can be omniscient."[105] Continuing to write for Rome, he invented in 1942 the fictitious characters Piero Mazda and American Imperialist, read by the voices of Radio Rome. Later he contributed brief items for a feature known as "News from Nowhere," which was dedicated "to ridiculing English news broadcasts."[106] After Italian fascists founded the short-lived Salò Republic in 1944, Pound continued to contribute speeches and sketches to Radio Milan: "From about May to September 1944, I sent items to the Republican Fascist Radio at Milano. When I wanted my name known or used, these items were written in the form of an interview."[107]

While broadcasting for Radio Rome, Pound's duties also included his participation in "Round the Mike" sketches, in which two or three announcers gathered around the microphone and conversed. Interestingly, this sometimes led to him being on the air without his knowledge: "I spoke to you, I thought," Pound says in one transcription. "I was not talking to anybody but the boys around the microphone."[108] Indeed, being overheard "round the mike" was one of the few occasions in which Pound could be heard "live" over the air: "At first, for a very brief time, I used to speak directly over the Air, but on one occasion during 1940 I made some remarks at the end of my talk, not in the script, simply a repeat of a main point, and after that incident I was ordered by PARESCE to record my talks on a disk, and this disk would be rebroadcast over the Air."[109] Pound's voice was stored on disks for later broadcast so as to make certain he would not express anything untoward of the Fascist authorities; they suspected, much like the American camp censor who objected to the *Pisan Cantos,* that Pound was inserting hidden messages into the transmission.[110] In fact, once on the air Pound was prone to improvisation, so connecting the microphone to the master disk could allow for the requisite security checks. Pound's ostensibly live broadcasts were not live.

We ought not dwell exclusively on this testimony, however, as other examples of his more nuanced approach to the importance of voice storage, specifically with regard to sequential manipulation, are available. In a letter to Ranieri, dated June 1, 1941, Pound expressed the importance of voice sequence: "The sequence of discs is important, at least to me, on supposition that ANYone listens and that there IS any mental sequence in the U.S. AT ALL. My stuff has sequence, i.e. idea suggested, idea repeated, idea."[111] Pound recognized that the sequence of his recordings, transmitted wirelessly, needed a listening "ANYone" in order to register the sequence. His formula condenses, to an astonishing degree, the relationship between manipulation and effect into programming his listeners for movement. More details may be found in a letter Pound received in 1939 from Jim Angleton, cited by the poet's daughter as key to reviving Pound's interest in radio:

> Maybe this will interest you, MacLeish is the innovator. . . . The idea is that
> every American has a couple of ears and that the ear is half poet. That by

radio a vast crowd is reached which gets the muse by flicking a button. Hence whole masses can hear. . . . The poet chooses social subjects and whatever he pleases . . . the only part that we are concerned with is broadcasting on records and rebroadcasting until the proper effect has been attained.[112]

Angleton's letter demonstrates an informed appreciation of the technological materialities of poetry in a radio age. He perceives that an entire nation can revel in the *jouissance* of the ear and that the muse could be contacted not through the previous century's slow and inefficient handwritten communication technology but merely by touching a button. This occurs because the wireless-gramophone linkup opens the one ear when the broadcasting and rebroadcasting of stored voices attain their "proper effect." In different words, wireless broadcasting of records supplies the data that a listener's ear has been programmed to hear as poetry.

One also notes the ease with which Angleton elides the subject of the poet's talk since any acoustic data will work to attain the proper effect as long as they are on record, that is, are capable of being manipulated (broadcast and rebroadcast). Left unsaid in Angleton's letter is what happens when the vast crowd "gets the muse." Before technological voice storage and transmission, getting the muse would have referred to the command to dictation a writing hand took from a Romantic imagination. In a radio age, the command has been transferred to a consumer of sounds who can, with a flick of a button, turn the muse on and off. Thus, the hookup becomes a three-in-one affair: a source of poetry, a transmission of poetry, and a command that it be heard. In its essential outlines, the radio brings together data, addresses, and commands in itself, what before had operated under the domain of writing.[113] A poetry in direct competition with a device capable of joining together in the broadcast of a single sound the three functions of an information system will attempt the same. Pound's search across various media for the ideogram, hieroglyphic, word, or sound that combined data, address, and command protocols in itself is the proof, demonstrating that his output in writing and over the wireless ought to be read in just these terms. Hence the importance that Angleton's letter merits.

Dictation at the Typewriter

Angleton and Pound's insight into disk storage confirms one of the wireless's principal lessons: garnering attention for a transmission moves through the differentiation of voices that the wireless-gramophone hookup makes possible. The poet broadcasting to the poetic ear of a crowd will attain an effect through the manipulation of his recorded voice; certainly Pound intended effects when he devised a playlist of his own speeches. Yet, Pound's speeches were not simply to be broadcast; they were to be published as "300 Radio Speeches," a reminder that the disks storing Pound's voice are the other of his typed copy.[114] Put differently, the relation between Pound and his typed notes conditioned the hookup between disk and transmission. In the last chapter, I described the role the typewriter and Pound's core group of secretaries had for the composition of the *Pisan Cantos*. The same relation operates in the production of written speeches for the radio, and although there are significant omissions in the accounts, there is enough information to speculate on the relation between sound transmissions and dictations a typewriting daughter takes down.

That the speeches were written wirelessly and not simply for the wireless Pound himself admitted. "My notes for Rome radio and my thoughts are telegraphic . . . agenda or agendum, more use than discussion. You might even have it marconigrammed (telegraphed)."[115] As this study has repeatedly demonstrated, there is no "marconigrammed" message without a vibrating Marconi nearby (or inside) taking dictation. Pound's notes that could have been signaled via voice or tapped transmission are the result of a wireless interface of a poet's brain and an instrument that isolates sounds, readying them for transmission. The typewriter provides Pound with a supply of signifiers through which streams of data move. Under the right media-technological conditions, Pound occupies the relay station of an immense wireless network.

A first requirement is the capacity to gather information: "He gathered news and information from Italian newspapers and whatever foreign papers he managed to obtain; from Italian broadcasts and any foreign station (especially the BBC) he could hear on his own radio; from conversations with friends in America and other countries; and from his own library, which included back numbers of periodicals."[116] The reader will recall Pound's association of poetry to data collection, which I outlined in the previous chapter. There I showed how Pound maximized the dispatch potential of his poetry by switching between voices. Here Pound is the interface between differently formatted data that a wireless-gramophone linkup alone does not process; not because it was impossible, but because Pound was still writing. The data of printed matter, voices, radio broadcasts, and of course Pound's *Cantos* require the proper means with which to store, retrieve, filter, classify, distribute and display them.[117] The typewriter occupies an important role, for by isolating letters and words, it rewards sparseness over loquacity and readies them as marconigrams. "Writing as keystrokes, spacing, and the automatics of discrete block letters bypassed a whole system of education," Kittler writes.[118] In Pound's case, they also implemented a coupling between wireless transmissions and poetry.

Two typewriters take dictation to produce the radioscripts. In the first, the artist-antenna dictates necessary information to a portable Corona typewriter, routing its messages through Pound's body, turning him into a medium. Perhaps this is the dictation Pound's court-appointed psychologist speaks of when describing Pound's mental state at the time of the transmissions.

> Q: Does he have any loss of memory as to the fact that he did write manuscripts and broadcast them?
>
> A: He speaks of that. His memory on some things is quite uncertain. There was some discussion, I remember, of him having dictated some manuscripts for broadcasts.[119]

But one typewriter is unequal to the task of processing various data streams. The poet writing under conditions of wireless media requires female typists, literally typewriters, to manage data flows. In Pound's case his daughter chiefly operated these typewriters.[120] She received his dictation, and offered her hands, ears, and mouth in an early example of word processing.

> In place of the bunch of letters he read us his radio speeches. We were his first audience, and, in the light of what we read in the papers and heard over the Italian radio, they seemed to me clear and justified. . . . After reading to us what he had written, it was my turn: I had to recite five lines from the *Odyssey* and translate. . . . And it seemed as though he possessed two voices: one angry, sardonic, sometimes shrill and violent for the radio speeches; one calm, harmonious, heroic for Homer, as though he were taking a deep, refreshing plunge into the wine-colored sea after a scorching battle.[121]

The substitution of letters for speeches marks a quaint scene of familial broadcasting, whose veracity will be confirmed in light of the print and broadcasts that have been inscribed in a secretary's brain. In addition, Pound's transmission is met by another, the recitation and translation of lines learned by heart, which is exactly what Pound's radioscript demands of its listeners: secretaries who can take dictation, recite from memory and translate, confirming the suspicion that Pound does not want to be read but to be learned by heart.[122]

As long as his secretary lacks a typewriter, however, she is simply the target of "blank shots." Accompanying her father to a meeting with Santayana, Mary de Rachewiltz writes:

> "No philosophy until you are forty" was the refrain, but on our way I sometimes had the impression that he was using me as a sort of blank wall to shoot his thoughts at, like tennis balls, in an attempt to put his ideas in order, round up his sentences, and hit the bulls-eye. I understood little or nothing of what he said; there were names I had never heard and concepts that resembled religious speculations. Since he still had to speak in Italian when trying to make me understand something, he groped for words as though verifying a coin from both sides.[123]

Before her introduction to the typewriter, the soon-to-be manager of Pound's discourse is the surface that registers the accuracy of his shots, a background on which Pound's verbal tennis balls strike, creating rounded sentences and ideas properly ordered. Thoughts hit blank walls; understanding little or nothing enables her to pick out names never heard and possibly even religious speculations.[124] She guarantees the transparency that allows Pound to verify his coins, the writing surface on which he will set his signs. Her banishment from the philosophical realm until age forty also serves to keep her from obeying the voice of hermeneutics to jump to the signifieds; she captures the command to inscription on the first page of her *Discretions*: "For years I have resisted a voice: 'Write it down, write it down!'"[125] In return she will be inscribed with signifiers that prepare her for proper dictation.

This scene of transmission indicates how indebted wireless writing is to the inscription a typewriter enables. Both install blindness in their acts of writing so as to better impress their words into memory, which, as our tour through Sergi's experiments in chapter 3 showed, occurs by torturing ears. Essentially, the process concerns embedding sounds: "Their sound [the *Cantos*], the way Babbo had so often read them in Venice, without my having understood a word, was somehow embedded in me, something very harmonious and beautiful. But reading them myself came as a shock. . . . Perhaps if I had read the first Canto first, things might have fallen into place earlier, but a consecutive reading was discouraged, mainly because of my ignorance and limited English."[126] Exactly. Unlike scores of Pound's readers steeped in consecutive reading, a budding secretary has no need for understanding them because only tennis-ball signifiers strike the blank wall.

Pound continues these transmissions later, even as her formal education provides her with a more orderly learning. He provided "action, work, in streams and in flashes." And by the logic of maximum energy in minimum signs, "what carried weight in my life then was Babbo's inner order; everything would forever depend upon that."[127] Of course, poetry is the medium of weight and not prose: "Many shades of emotion will remain hidden, embedded in the Cantos as mythology, since poetry is the true medium for truth. Prose fits facts, but facts carry little weight, they can simply be recorded as part of 'the tale of the tribe.'"[128] Once embedded, the registered sounds command attention, functioning as a medium for truth to emerge.

It is precisely the ability of the inscribed sound to draw attention to itself that produces the pain necessary for truth under its assorted forms to appear: it arrives in a flash via its isolation from other sounds. These forms Pound identifies in a brilliant article from 1924, in which he accounts for the success of modern music in rallying attention to itself. After detailing the sensory overload used by Wagner to create receptivity (an ideal not so far removed from the experiments of the "Radio *Cantos*"), Pound urges a second path that will utilize pain to intensify awareness:

> The other aesthetic has been approved by Brancusi, Lewis, the vorticist manifestos; it aims at focussing the mind on a given definition of form, or rhythm, so intensely that it becomes not only more aware of that given form, but more sensitive to all other forms, rhythms, defined planes, or masses.
>
> It is a scaling of eye-balls, a castigating or purging of aural cortices; a sharpening of verbal apperceptions. It is by no means an emollient.[129]

One will find no better summary of a wireless aesthetic than this. The sound inscribed by the gramophone and isolated by the artificial cuts of a wireless engineer or poet focuses the mind to such a degree that not only attention to form arises but an increased sensitivity to other forms results. Pound's statement confirms Arnheim's thesis regarding the radio: attention creates movement. The efficacious, inscribed sound is one that punishes aural cortices, causing pain sufficient enough to be classified as a nonemollient.

In the case of the daughter-father media link, the dictated word a typist lacking instruction in office technology hears and does not understand reaches into her brain and embeds itself. Hearing the embedded sounds of the *Cantos* brings with it a command to write them down; copying, translating, and editing his poetry continually brings Pound's discourse back to him. But a secretary pays in purged aural labyrinths, which transforms her into a gramophone: "As though placing the needle on a record at random—well knowing the beginning and the end—words came to the surface: *Il nonno* [grandfather] wants to see you again."[130] Naturally, she obeys. "Underneath all my theological dilemmas and dreams of heroism, I listened attentively to the summons: learn how to write. I was a slow thinker. How does one go about 'thinking'—thinking as different from describing, remembering, inventing. The question seemed too complicated."[131] Whereupon she cites her father's instruction to practice "strokes" as in tennis. Truth emerges automatically in the keyed strokes of a typewriter. Hitting the keys unconsciously spells out truth.

In the wireless dictations running from father to daughter, writing's traditional parameters—describing, remembering, and inventing—remain the prerogative of the poet/father. Once a secretary learns how to type, the *key* strokes that inscribe signs into memory will make taking dictation easier. Not coincidentally, her "writing" begins with the discovery of Pound's discarded typewriters:

> Having taught myself to type with two fingers on one of Babbo's discarded old portable Corona typewriters, I cherished the childish notion that once a page looked neat one should not touch it again, but Babbo would always

change around a word, scribble in an alternative: try it out. My slow, careful typing all messed up again.[132]

As she picks up speed, the more she takes down. "Three volumes on the life and work of the troubadours became light reading. The more I got absorbed in the Cantos the more eager I became for further knowledge."[133] Her eagerness, I repeat, results from making automatic the keystrokes that strike the blank page. This eagerness in a radio age comes about when sounds have been sufficiently isolated so as to purge aural passages and embed sounds, calling her to attention. In a wireless network, the fast typist is transformed into a listener who takes dictation without a typewriter. Automatic strokes result in "words that stick in my mind."[134] Thus, we have another example of media linking up and achieving dominance—the typewriter helps determine the mode of radio reception and transmission. Pound does not transmit meaning in his dictations to his daughter but only the materiality of the medium in question. When listening to a broadcast, it is the embedded sounds that issue commands to write. Pound issues the order formerly given to his daughter in his "E. E. Cummings" broadcast: "I don't suppose you stenografize, but you better take general notes on what I am tellin you other wise there is going to be a tall lot missin!"[135] What matters is the difference between letters or sounds that become visible or audible when bounced off a blank page or a blank wall.

Writing's Protocols

Secretaries are indispensable for a fuller accounting of Pound's poetic production. They are neither transmitters nor receivers so much as relays that a partially connected media system utilizes in order to commit words to paper (or the airwaves), be the typewriters machines or daughters.[136] Their major effect is to ready Pound's poetry for its proper coupling with appropriate transmission devices: writing published in book format or, during World War II, as voices stored on disks for later transmission. Yet in the latter a new element has been added. Writing can no longer count on the series of black letters on white pages for its sequence, that is, wireless writing does not simply store itself as it had until then but depends on another medium for its sequence. As radical as Pound's experiments in voice switching are in the "Radio *Cantos*," in his speeches over the radio, his writing competes primarily with the wireless-gramophone hookup. When Pound finally gains access to a microphone that records his voice onto disks, a wirelessly inflected writing loses its grounding in the linearity lines provide. Rather than utilizing typography for modulating frequencies, Pound is forced into using contiguous sounds in order to create meaning.

To create meaning with contiguous sounds, Pound modulates the frequencies of his voice in order to set off one tone from another. Following Arnheim, he must give "the whole situation" using only the materials at hand; he ventriloquizes, shifting among voices, superimposing sounds, giving his auditors the sensory evidence to relate each sound to another. Here Pound was aided immeasurably by his way of speaking: "His way of talking was just as odd as his way of typing. He spoke with many voices. In the midst of expositions in a flat, pedantic, and occasionally scolding tone, he would lapse into exaggerated Western drawls, Yankee twangings, feet-on-the-cracker-barrel pipings, and as suddenly switch to upper-class British sibilants and even Cockney growls. But while he looked and sometimes sounded like a mountebank, he was in deadly earnest."[137] Or as another of Pound's biographers writes: "He spoke in a variety of accents, American and English, grunted and exclaimed, and sometimes jumped about from point to point in such a way that he must have been difficult to follow."[138]

What is the relationship between the variety of accents Pound uses and his jumps from point to point? Anyone reading Pound's original manuscripts (which Doob transcribes) will find the broadcasts' most salient characteristic missing: the switches that signal frequency modulation between words and sentences are not written down, as typewriting cannot provide the real components of voice. Pound's written notations for pronunciation, the capitalizations and abbreviations that he uses to indicate the necessary modulation, are surely a reflection of their broadcast status, but these strategies remain within the orbit of the Gutenberg Galaxy: another medium liquidates the claim of the written page to storage. And with the uncoupling of one of writing's subroutines, a "writing minus grammar, orthography and the norm of the high idiom" results.[139] Remove storage from writing's protocols and Pound is commanded to write as he speaks, prompted to match the speed with which disks inscribe his voice. Like many before him, Pound becomes a mere relay for the embedded sounds of disk storage. The media system utilizes storage to draw out a speech that the wireless transmits, after which the speech is fed back to Pound who, as an avid listener of his broadcasts, hears himself elsewhere. In true feedback fashion, precision results, new associations are created and truth is heard. Pound captures these consequences in his radioscript "Zion": "Is it possible to arouse any interest or verbal precision . . . Cabala, for example, anything to make the word mean something it does not SAY. Anything to distract the auditor from the plain sense of the word, or the sentence."[140]

Pound's broadcast from July 6, 1942, "Continuity," is a fine example of what occurs when acoustic devices spell out sequence. The speech "begins" when Pound makes the delirium of transmission the necessary ground for what is written. "Had I the tongue of men and angels, I should be unable to make sure even the most faithful listeners would be able to hear and grasp the whole of a series of my talks."[141] The difficulty of speaking, he explains, inheres in radio form, and although not explicitly identified, we are assured the problem concerns sequence: "Nevertheless you may as well make the effort to grasp at least the fact that there IS a sequence in what I am saying, and that the conversation of February coheres with that of April."[142] Where the radio fails to provide proper addresses for its transmissions, books give exact addresses in chapters and pages, so "that [the] reader CAN, when he wishes, look back take up the statement of the preface, see where Chapter X hitches onto Chapter I." What remains in an environment purged of specific addresses is the hope that "you . . . get [the] main threads and cables of what I am telling you." This is because cables belong to the world of telegraphs and do not take account of the specific properties of how sounds "mean" when transmitted over the wireless.

When sequence becomes a function of voice storage, the temporal coordinates of beginnings and endings shift because the speed of inscription that conditions playback speed also allows for manipulation. So Pound has trouble commencing: "And after a hundred broadcasts it is STILL hard to know where to begin. There is so MUCH that the United States does not know."[143] Pound's inscribed words cannot simply call up an optical signifier on the written page to provide the words with a duration not tied to their moment of sound on a record. By the time Pound makes his broadcasts, manipulation and storage coincide: he could be played back at different speeds, shifting time to meet the exigencies of transmission. It is not just a case of Pound jumping from point to point but rather of him registering what storage technology made possible. Thus, the possibility of time-axis reversal becomes real, as Arnheim had indicated. A wireless engineer could quite easily augment the speed of the sound strip or disk or even retrace the recording's vibrations from the end to the beginning; speech can be played backward.

Pound's speech from July 6, 1942, works along these lines, while remaining within the limits of (type)writing. Consider the ease with which one can read the speech backward, not word for word as in a massive palindrome, but on the level of sentence and paragraph. There is little logical connection between Pound's assertions, outside of an occasional "for example" when he speaks of Manet and rare references to before and after. The overall effect of *reading* the speech is of nonmovement, which Pound explicitly recognizes in his closing: "I know I haven't gotten very far in this talk, so wait for the next one." The paragraph starting "I know all about the chocolate box" is exemplary. One could begin with the paragraph's last sentence, "Early drawings, Burne-Jones, that's what he WANTED," and, continuing in this vein, read to the beginning, and the "sense" of the paragraph would not be different. Pound does not get far because, when writing under conditions of time reversal, the written no longer anchors its own playback. Pound gestures to time reversal and manipulation off-handedly: perhaps that is the reason he includes his customary opening signature, "Ezra Pound speaking," in the penultimate paragraph of the speech, suggesting both an understanding of the new demands of radio (someone might be just tuning in) and the possible reversal of the speech's sequence.[144]

The "Pell-Mell" Registration of Wireless Dictation

Just as the typewriter played a crucial role in accounting for the wireless format of *Guide to Kulchur,* so too does it help determine modes of information processing in Pound's voice recordings. Here processing occurs in conjunction with other media. First, Pound must find a data source. Some we have already surveyed—print media, the BBC, other Axis broadcasts, Pound's own contributions to the *American Hour* and the *Cantos* themselves—but unlike other wireless writers, Pound has no need of amphetamines, naval vessels armed with transmitters, or positivists torturing him to recognize spaces (not yet at least).[145] Instead he has at his fingertips a device that transmits data. A radio gets him started as a sound collector, allowing him to turn the dial, as he mentions in his talk from March 7, 1943: "And now in the year 1943 I find it almost impossible to listen to London. I stuck along a while with your social comedy. . . . But now I just wearily turn the button."[146]

Next Pound requires a device with which to type the messages he has received, using ribbons to inscribe them on paper. He immediately faces two difficulties. First, the amount of information under wartime has increased dramatically. This was clearly less of a problem when data storage was limited to writing—witness Pound's experience during World War I. Unable or unwilling to make it to the front, Pound remained at his typewriter: "[I]t is ironical that I who care something for civilization should be left at my typewriter."[147] When the recording and manipulation of speech become standard, however, Pound has trouble keeping up with the demands the radio makes on his capacity to process data. Ideally, had Pound transmitted directly and automatically to the microphone and over the air, without passing through the letters and spacing of a typewriter, he might have been better able to pass along, if not all, then certainly the crucial components of the information he was picking up.[148] He could have turned himself into the medium of a medium, à la Marinetti and the wireless; rather than finding a hand disconnected from a brain, one would have found Pound running at the mouth. Second, for a poet whose writing was already wirelessly inflected, further condensation in recorded voice transmissions is only remotely possible. Had Pound simply been writing and not broadcasting, he would surely have followed the strategy, in *Rock-Drill* and elsewhere, of exploiting typography to transmit more information. As it is, the wireless voice transmissions that Pound writes are not all equally up to the demands of efficient information processing.

Pound's radio broadcasts ought to be read with this optic: in the struggle of the wireless network to reach proper connectivity among its various media components (Pound's brain one among a number), a transmission records its failure to process the variously formatted data of writing, other broadcasts, and memory. Thus, Pound types and broadcasts in his speech of July 6, 1942: "I am held up, enraged, by the delay needed to change a typing ribbon, so much is there that OUGHT to be put into the young American head. Don't know which, what to put down, can't write two scripts at once. NECESSARY facts, ideas, come in pell-mell. I try to get too much into ten minutes. Condensed form O.K. in book, saves eyesight, reader can turn back, can look at a summary."[149] Note Pound's failure to keep pace with the acceleration of information movement created by the linkup with the wireless-gramophone. His messages hypertrophy since old media are unable to manage information that require new media; we of course read the effects. When information levels and speeds rise, the outlines of the limits of condensation become clear. In book format, eyesight is saved, *Guide to Kulchur* may be written, but little can be processed given the speed necessary to capture the "pell-mell" flows of facts and ideas. Pound cannot write two scripts at once, which confirms Arnheim's argument regarding the nature of wireless sound processing. By extending the different spatial characteristics of sounding bodies in depth, the wireless transmission works nonlinearly. It would seem that Pound is impelled to attain a different, more overall knowledge of process than the one he is utilizing here.

Perhaps, the reader will recall Pound's use of weightless ideograms in *Guide to Kulchur* to inscribe sound events into memory. There was no use in retaining knowledge "in the form of dead catalogues once you understand process."[150] But here Pound fails to adopt a stance that would have allowed him to keep up. Pound's dilemma is one McLuhan analyzes in a forgotten article from 1960. "Sequential analysis and adjustment natural to low speed information movement becomes irrelevant and useless even at telegraph speed. But as speed increases, the understanding of process in all kinds of structures and situations becomes relatively simple. We can literally escape into understanding when the patterns of process become manifest."[151] Pound's incapacity to "escape into understanding" occurs despite his numerous broadcast pleas to the contrary: "PITY is that there is so much else, so much essential that they are unblissfully UNaware of. And I honestly do not know where they can get essential parts of that else, except from my broadcasts."[152] In purely media terms, Pound does perform many operations at the same instant as part of the information flow: he accesses the data, he receives it, and then he stores it. The difficulty concerns the connection to the typewriter and the difficulty keeping up with the accelerated information movement created by the introduction of the interface with the wireless network or, more precisely, the wireless superimposition of sounds. The wireless-gramophone hookup works nonsequentially and instantaneously—its lack of segmentation, when broadcast to a poet taking dictation, becomes the stuff of lost threads. This is because the typewriter remains within the Gutenberg Galaxy, guaranteeing uniformity and repeatability in sequences. Pound's broadcasts therefore measure the struggle between writing and wireless to register data and mobilize listeners.

To be sure, this summary of Pound's production for Radio Rome is not without paradox, which is to be expected considering the various media technologies that made them possible. On the one hand, the use of sound recording isolated and intensified components of Pound's voice, essentially doing the work of the ear and all that implies. Pound's attention-gathering method of switching among voices transposed into writing features of gramophonic inscription in order to hit on the registering area of the brain. On the other hand, changes in transmission technology accelerated data flows to a typewriting Pound (and daughter), what an early student of radio noted thusly: "It offers, all but forces on us, new and wider horizons of knowledge or, perhaps I should say, of information."[153] Pound's failure to manage the data flows that circulated between him and his writing machines does not diminish the lessons of media competition for transmitting poets. The chief lesson? The materiality of the written is often realized at the expense of being understood over the radio. Or as another more recent student of radio has remarked: "What is at stake is the materially constituted relationship and not understanding per se or direct, nontransformational or regularly transformational inputs from the 'world' to consciousness."[154]

The Feedback Loops of Fascism

At this juncture, we are in position to return to some of the knottier questions that presented themselves in our appropriation of Arnheim for a reading of Pound's wireless aesthetics. In order of importance: How do apparently incoherent transmissions register with listeners and so create movement? Are sparse signifiers and sound cut-ups that release energy already on the road to fascism? How might we think together wireless information processing with the larger fascist project of "enhanced individual autonomy" and "enriched human agency" in part through technology?[155] Responding to the first question, we recall that successful wireless transmissions are based not on sense so much as creating movement by means of sound superimposition. Recognizing the installation of blindness that accompanies wireless transmissions, poets on the blank page or over shortwave become wireless engineers, practicing sound cutting and frequency modulation on their own bodies.

Moreover, modulating frequencies and exploiting the possibilities of time-axis manipulation embed sounds in that part of the listener's brain that is capable of registering; proof, if we needed it, that a genealogy of wireless technology must be first sought in the neuropsychology and positivist sciences that measure human responses. Embedded sounds, in turn, function as media for other commands to arise: in the case of Pound's daughter, the command was to write down, but other commands are equally possible. The operation occurs by way of a transmission of energy that concretizes the formation of a subject precisely in the spacing of absence/presence. In *Guide to Kulchur,* a transmission was successful when assertions not connected logically among themselves began to carry weight in the fashion of a Boolean motor. The succession of signs that reached the reader's senses, if not making sense, did function as a switch that set movement in motion. Pound's broadcasts over Radio Rome attempt the same, within the confines of an outdated media linkup that uses a typewriter to connect its speaker to the gramophone. Quite possibly, it may be on account of the incomprehensibility of modulating frequencies in writing that certain words register in a reader's brain.

Is there an inherent association to be made between this process of wireless information processing, with its sound cuts and minimum signs giving off maximum energy, and fascism? Marinetti's use of *parole in libertà* to create short-circuits to *il duce* ought to be recalled in this regard. Installed wireless imaginations created telepathic relays between public and poet; the wireless command to register was expressed in the half-word or gesture that revealed truth to the budding Futurist who paid with his blood. Pound's radio broadcasts represent an advance on Marinetti's primitive hookup. Yes, fascism and the wireless remain coterminous in his broadcasts and *Guide to Kulchur* to the degree revelation is always waiting to be decoded when the wireless is turned on. But Pound's encounter with the wireless demonstrates more clearly than Marinetti's the nature of the feedback loop that runs between the sound manipulation made possible by recording technology and "codes of action." Only those listeners whose attention has been captured by frequency modulation are able to move in the direction indicated by the succession of signs.

This is the case whether the medium in question is poetry or, more formally, Pound's radio broadcasts made during World War II. For the wireless loop to close, not merely sense must be made but programming to create action must also commence. Pound's broadcasts, it bears repeating, do not require listeners trained in the hermeneutic arts but rather only surfaces (or daughters) to receive the sounds of his voice in motion. Voices pulsating across the ether have nothing to do with an intention on the part of Pound but reflect instead the nature of the medium. In the imperialism of the one voice installed by the wireless, even the noise emanating from the body may be sent. Pound's transmissions, therefore, model the dynamics of the wireless system by utilizing its feature of ubiquitous movement, sparse sound cuts, and purged aural cortices to reach listeners who, by taking down dictation, are programmed to move. A feedback loop is created between attentive listeners and those able to carry out the execution of the program: "kulchured" men and women reached by all the means at Pound's disposal.

Finally, how might we best associate the wireless characteristics of Pound's broadcasts and writing with the larger project of fascism/modernism? Leaving aside the sterile questions of whether Pound was a fascist (he was), and if so, how much of a fascist was he (very), the question essentially concerns whether fascism, like Pound's poetry, might be profitably viewed as a subsystem of a wireless network, with its own information processing and human interfaces. Lawrence Rainey's discussion of fascism and its inscription of the body politic is suggestive. Speaking of photographic postcards that show hundreds of children arranged to form letters, he argues that fascism issued a new syntax, a somatic language "that cannot be deciphered by the citizens who have been, quite literally, conscripted into its making." Only the master aviator, he who has dictated the script, is able to decipher it from high above.[156] Rainey's argument is that once deciphered, these letters point paradoxically to enhanced individual autonomy and human agency along with an enthrallment to technology.

The wireless confirms and extends Rainey's insight by focusing our attention on constant movement, what Mussolini repeated in his speeches: "Fascism is movement, not stasis. It is constant combat, not barren waiting."[157] The wireless is the device that typifies Mussolini's definition, for it continually transposes static into movement, addressing transmissions to ears and introducing an information processing that relies on blindness to conscript bodies into movement. Thus Rainey's insight into syntaxes that conscript and inscribe bodies traces the same route traversed in our examination of Pound's broadcasts and writing. It comes down to dictation. In the broken feedback loops of consciousness created by the wireless-gramophone hookup, the wireless is the announcer or dictator who speaks but does not hear himself, while the rest take dictation without being able to speak. Paraphrasing a functionary of the Germany Propaganda Ministry, the fascist is he who, in taking dictation from the wireless, knows how to spell radio correctly, "with three exclamation marks because we are possessed in it of a miraculous power—the strongest weapon ever given to the spirit—that opens hearts and does not stop at the borders of cities and does not turn back before closed doors; that jumps rivers, mountains, and seas; that is able to force peoples under the spell of one powerful spirit."[158]

NOTES

INTRODUCTION

1. Marshall McLuhan, *Understanding Media: The Extensions of Man* (New York: McGraw-Hill, 1967), 305.

2. On Marinetti's deferred engagement with radio in the interwar period, see Margaret Fisher, *Ezra Pound's Radio Operas: The BBC Experiments, 1931–1933* (Cambridge, MA: MIT Press, 2002), 45–54.

3. Until recently, Hugh G. J. Aitken's *Syntony and Spark: The Origins of Radio* (Princeton, NJ: Princeton University Press, 1985), esp. 298–336, was the best history of the wireless and the scientific research that preceded the development of wireless technology. Sungook Hong's more recent study, *Wireless: From Marconi's Black-Box to the Audion* (Cambridge, MA: MIT Press, 2001), now deserves that honor for its brilliant reading of the wireless and audion. Although I part ways with Hong, especially concerning the greater weight I afford the superheterodyne in the period's later media ecology, his study is the best to date of the specifications of early wireless technology. For a more general account, the reader is directed to Tony Devereux's *Messenger Gods of Battle: Radio, Radar, Sonar; The Story of Electronics in War* (London: Brassey's, 1991) and those sections devoted to functioning of the wireless in Gavin Weightman's *Signor Marconi's Magic Box: How an Amateur Inventor Defied Scientists and Began the Radio Revolution* (London: HarperCollins, 2003).

4. The history of wireless invention remains contested. Some, notably Friedrich Kittler, insist on the role of Professor Adolf Slaby, while others, including Margaret Cheney, award Nikola Tesla the distinction of having first transmitted wirelessly; still others argue for British scientist David Edward Hughes. Thus, writing in February 1901, Tesla claimed to have invented his own system of telegraphy without wires as early as 1893, two years before Marconi. Hughes, for his part, had apparently constructed a portable wireless transmitter in London as early as 1879, almost a decade before the discovery of Hertzian waves. Without rehearsing the various claims to the wireless mantle here, what seems clear is that a number of figures contributed decisively to the emergence of the wireless before Marconi: Heinrich Hertz's monumental discovery in 1888 that sparks could be employed to generate electromagnetic waves (soon known as Hertzian waves); Edouard Branly's experiments in 1890 demonstrating the effects of electric charges on metal filings in test tubes; and Oliver Lodge's proof that Branly's tube could detect Hertzian waves. All of which is to say that Marconi was not the first

171

to transmit wirelessly. Crucially, however, Marconi did extend the range of the wireless transmission well beyond his predecessors, transmitting wirelessly across the Atlantic in 1901, while, perhaps more important, successfully marketing his wireless to British scientific authorities. See in this regard Weightman, *Signor Marconi's Magic Box,* especially pages 93–99, from which I have drawn here. For Slaby's claims, see Friedrich A. Kittler, *Gramophone, Film, Typewriter,* trans. Geoffrey Winthrop-Young and Michael Wutz (Stanford, CA: Stanford University Press, 1999), 94–96. Margaret Cheney presents her case for Tesla in *Nikola Tesla: Man Out of Time* (Englewood Cliffs, NJ: Prentice Hall, 1981). For general histories of wireless invention, Aitken's *Syntony and Spark* offers an excellent account of the wireless and the scientific research that preceded its development. Hong brilliantly reads the specifications of early wireless technology and individual contributions in *Wireless.*

5. Aitken, *Syntony and Spark,* 192–93.

6. "The first documented American radio broadcast was an experiment conducted by R. A. Fessenden from Brant Rock, Massachusetts, to ships at sea on December 24, 1906. Fessenden's program included a short talk, the singing of carols, a work by Handel, and recitation of a poem" (Aitken, *Syntony and Spark,* 69, 74; quoted in Fisher, *Ezra Pound's Radio Operas,* 41). Cf. Adorno on musical reproduction over radio: "Radio is essentially a new technique of musical reproduction." Theodor Adorno, "A Social Critique of Radio," in *Radiotext(e),* ed. Neil Strauss (New York: Semiotext(e), 1993), 275.

7. Rudolf Arnheim, *Radio,* trans. Margaret Ludwig and Herbert Read (London: Faber & Faber, 1936), 128.

8. Hadley Cantril and Gordon W. Allport, *The Psychology of Radio* (New York: Harper & Brothers, 1935), 18.

9. Gaston Bachelard, "Reverie and Radio," *Radiotext(e),* 219, 218.

10. R. Murray Schafer, *The Soundscape: Our Sonic Environment and the Tuning of the World* (1977; Rochester, VT: Destiny Books, 1994), 215. Yet Schafer's notion of the information processing of sound gestures to some of the same systemic concerns I consider here. "Second, broadcasting is a deliberate attempt to regulate the flow of information according to human responses and information-processing capabilities" (234).

11. Susan M. Squier, ed., *Communities of the Air: Radio Century, Radio Culture* (Durham, NC: Duke University Press, 2003). See also Jeffrey Sconce, *Haunted Media: Electronic Presence from Telegraphy to Television* (Durham, NC: Duke University Press, 2000), and Tom Lewis, *Empire of the Air: The Men Who Made Radio* (New York: Edward Burlingame Books, 1991).

12. Jacques Derrida, *The Ear of the Other: Otiobiography, Transference, Translation,* trans. Peggy Kamuf (Lincoln: University of Nebraska Press, 1985), 28.

13. Cf. Derrida's notion of interiorization. "Right at first the sound touches us, interests us, impassions us all the more because it penetrates us. It is the element of interiority because its essence, its own energy, implies that its reception is obligatory." Jacques Derrida, *Of Grammatology,* trans. Gayatri Chakravorty Spivak, Corrected Edition (Baltimore, MD: Johns Hopkins University Press, 1974), 240.

14. "The legal fight to insure copyright, the cult of the author, print technology, and print culture worked hand in glove to create a depth model of subjectivity in which ana-

logue resemblances guaranteed that the surface of the page was matched by an imagined interior within the author, which evoked and also was produced by a similarly imagined interior in the reader." N. Katherine Hayles, "Simulating Narratives," *Critical Inquiry* 26, no. 1 (Autumn 1999): 14.

15. Douglas Kahn, *Wireless Imagination: Sound, Radio, and the Avant-Garde* (Cambridge, MA: MIT Press, 1994), 21.

16. John Johnston, "Mediality in *Vineland* and *Neuromancer*," in *Reading Matters: Narrative in the New Media Ecology,* ed. Joseph Tabbi and Michael Wutz (Ithaca, NY: Cornell University Press, 1997), 174. "Meanwhile, however, there are still media, and today's technological standard can be described in terms of 'partially connected media systems.'" See John Johnston, "Friedrich Kittler: Media Theory after Poststructuralism," in *Literature Media Information Systems: Friedrich A. Kittler Essays* (Amsterdam: Overseas Publishers Association, 1997), 3.

17. "The general digitization of channels and information erases the differences among individual media. Sound and image, voice and text are reduced to surface effects, known to consumers as interface." Kittler, *Gramophone, Film, Typewriter,* 1.

18. Ibid., 43.

19. For a summary of the events at Fiume, see chapter 2.

20. N. Katherine Hayles, *Writing Machines* (Cambridge, MA: MIT Press, 2002), 104.

21. Janet Cardiff's recent retrospective at the Castello di Rivoli in Turin deeply engages a wireless sonority (Janet Cardiff, *A Survey of Works Including Collaborations with George Bures Miller,* Turin, Italy, May 21–August 31, 2003). See also media artist Paul DeMarinis's recent explorations of the implications of twentieth- and twenty-first-century wireless voice transmissions. A sampling may be found at http://www.well.com/~demarini/installations.html.

1. Marconi, *Marconista*

1. F. T. Marinetti, "Destruction of Syntax—Wireless Imagination—Words-in-Freedom," in *Futurist Manifestos,* trans. Robert Brain, R. W. Flint, et al. (New York: Viking Press, 1970), 95–106; Vicente Huidobro, *Obras completas de Vicente Huidobro* (Santiago: Editorial Andres Bello, 1976); Gabriele D'Annunzio, *La penultima ventura: Scritti e discorsi fiumani,* ed. Renzo De Felice (Milan: Arnoldo Mondadori Editore, 1974); Ezra Pound, *Guide to Kulchur* (New York: New Directions, 1970), 210.

2. Friedrich Kittler, *Discourse Networks, 1800/1900,* trans. Michael Metteer (Stanford, CA: Stanford University Press, 1990).

3. Friedrich Kittler, *Gramophone, Film, Typewriter,* trans. Geoffrey Winthrop Young and Michael Wutz (Stanford, CA: Stanford University Press, 1999), 211.

4. Laura Otis, "The Other End of the Wire: Uncertainties of Organic and Telegraphic Communication," *Configurations* 9 (2001): 193.

5. Kittler, *Gramophone, Film, Typewriter,* 102; John Johnston, "Mediality in *Vineland* and *Neuromancer*," in *Reading Matters: Narrative in the New Media Ecology,* ed. Joseph Tabbi and Michael Wutz (Ithaca, NY: Cornell University Press, 1997), 173.

6. Mark Seltzer, *Bodies and Machines* (New York: Routledge, 1992), 12–13.

7. To my knowledge, only one Marconi biography appeared in the 1990s: Giancarlo

Masini's flawed *Guglielmo Marconi* (1975; New York: Marsilio Publishers, 1995). More helpful is W. J. Baker's *A History of the Marconi Company* (1970; London: Routledge, 1996), and one should not overlook Maria Marconi's recently reprinted love letter to her husband, *Marconi, My Beloved* (New York: Dante University of America Press, 1999). Other titles of interest include: Gavin Weightman, *Signor Marconi's Magic Box: How an Amateur Inventor Defied Scientists and Began the Radio Revolution* (London: HarperCollins, 2003); G. Falciasecca and B. Valotti, eds., *Guglielmo Marconi: Genio, Storia, Modernità* (Milan: Editoriale Giorgio Mondadori, 2003); G. R. M. Garratt, *The Early History of Radio: From Faraday to Marconi,* History of Technology Series 20 (London: Institution of Electrical Engineers, 1994); Peter R. Jensen, *Early Radio: In Marconi's Footsteps, 1894 to 1920* (Kenthurst, Australia: Kangaroo Press, 1994); and the first two chapters of Tom Lewis's *Empire of the Air: The Men Who Made Radio* (New York: Edward Burlingame Books, 1991).

8. Orrin Dunlap, *Marconi: The Man and His Wireless* (New York: Macmillan Company, 1937), 15. "Marconi not only added the antenna-land system that made the signaler of lightning bolts so efficient, but also added to it the Hertzian oscillator so as to augment its capacity, eventually equipping it with a telegraphic key to mark the emission of waves with the rhythm of the Morse alphabet." Franco Foresta Martin, *Dall'ambra alla radio* (Trieste: Editoriale Scienza, 1995), 112. Unless otherwise noted, all translations are my own.

9. Ibid., 16–17.

10. Ibid.

11. Jacques Derrida, *The Post Card: From Socrates to Freud and Beyond,* trans. Alan Bass (Chicago: University of Chicago Press, 1987); Jacques Lacan, "Seminar on 'The Purloined Letter,'" *Yale French Studies* 48 (1972): 38–72.

12. Briankle G. Chang, *Deconstructing Communication: Representation, Subject, and Economies of Knowledge* (Minneapolis: University of Minnesota Press, 1996), 47.

13. Ibid., 46.

14. I am grateful to Geoffrey Winthrop-Young for pointing this out. See his "Going Postal to Deliver Subjects: Remarks on a German Postal Apriori," *Angelaki* 7, no. 3 (2002): 143–58.

15. Luigi Solari, *Marconi: Dalla Borgata di Pontecchio a Sydney D'Australia* (Naples: Alberto Morano Editore, 1927), 64.

16. Adelmo Landini, *Marconi sulle vie dell'etere: La storica impresa narrata dall'Ufficiale Marconista dell'Elettra* (Turin: Società Editrice Internazionale, 1955), 22.

17. Despite its conflation of the telephone, radio, and wireless, Lynn Sharon Schwartz's account describes well the distilling effects: "Radio is a hardy art form, which has managed to stay alive and even thrive lately. . . . Its distinctive genius is to distill immediate presence into the voice—all gross matter purged away—and its best practitioners can make the voice, with its variables of tone, pitch, rhythm, and inflection, as rich a bearer of sensibility as silent dancers can make of the body." "Only Connect," in *Tolstoy's Dictaphone: Technology and the Muse,* ed. Sven Birkerts (St. Paul, MN: Graywolf Press, 1996), 129–30. On the symbolic significance of distillation, see Ernest Jones, "The Madonna's Conception Through the Ear," *Essays in Applied Psycho-Analysis*

(New York: International Universities Press, 1964), 320. "Beginning with the nutrient water, the source of all things, we see the coarser constituents being precipitated and discarded, while the finer elements, the essence of essences, are distilled over into vapour (pneuma), which in its turn is purified of any grossness still remaining and is rarefied into an aerial medium, ethereal and spiritual, intangible, invisible and indefinable–the psyche."

18. Degna Marconi, *My Father, Marconi* (New York: McGraw Hill, 1962), 107.

19. Anxiety was expressed in various ways. The British postal minister noted that the three dots signifying "S" and the dot-dash-dot indicating "R" in Morse code were those most often caused by these disturbances. One newspaper's account is typical: "Skepticism prevailed in the city. 'One swallow does not make a summer,' said one, 'and a series of "S" signals do not make the Morse code.' The view generally held was that electric strays and not rays were responsible for actuating the delicate recording instruments." *Daily Telegraph,* December 18, 1901; quoted in D. Marconi, *My Father,* 116.

20. Landini, *Marconi sulle vie dell'etere,* 45.

21. Ibid., 82.

22. "The quotidian data flow must be arrested before it can become image or sign. What is called style in art is only the switchboard of these scannings and selections." Kittler, *Gramophone, Film, Typewriter,* 104.

23. Landini, *Marconi sulle vie dell'etere,* 51–52.

24. "He is neither speaker nor writer. . . . And the larger sacrifice for him is writing. When he speaks and when he writes, he uses short and condensed phrases." Solari, *Marconi,* 45–46.

25. Cf. Kafka's operator in *America:* "In the words which he spoke into the mouthpiece he was very sparing and often one noticed that though he had some objections to raise or wishes to obtain more exact information, the next phrase he heard compelled him to lower his eyes and go on writing before he could carry out his intention." Quoted in Klaus Benesch, "Technology and the Failures of Representation in the Works of Franz Kafka," in *Reading Matters,* 88.

26. Johnston, "Mediality in *Vineland* and *Neuromancer,*" in *Reading Matters,* 17. The relevant section on modern myths of technological change may be found in Marshall McLuhan, *Understanding Media: The Extensions of Man* (New York: McGraw-Hill, 1967), 252.

27. For a recent example of the genre, see Weightman's *Signor Marconi's Magic Box.*

28. Landini, *Marconi sulle vie dell'etere,* 5.

29. Dunlap, *Marconi: The Man and His Wireless,* unnumbered.

30. Solari, *Marconi,* 5.

31. Ibid., 7.

32. Landini, *Marconi sulle vie dell'etere,* 14.

33. Ibid.

34. Ibid., 8.

35. Sigmund Freud, *Civilization and its Discontents,* trans. James Strachey (New York: W. W. Norton, 1961), 35.

36. "This is, in reality, the ultimate guarantee that the use of the category of the virtual

intends to offer man. A regression or a loss can always be considered virtually, a pro-gression or an acquisition. Hence the intrinsically consoling purpose that this cate-gory, which seems so abstract and speculative, proposes to achieve is evident." Nicola Abbagnano, *Critical Existentialism,* trans. Nino Langiulli (New York: Anchor Books, 1969), 108.

37. Following the argument, not just the superego is a technology. "In order to re-place the Id with an Ego, to replace violence with technology, it is necessary that one first fall into the clutches of this violence." Friedrich A. Kittler, "Dracula's Legacy," in *Litera-ture Media Information Systems,* ed. John Johnston (Amsterdam: Overseas Publishers Association, 1997), 56.

38. Landini, *Marconi sulle vie dell'etere,* 59.

39. Associating mothers to inventors and to a genus of imagined auditory recom-pense is of course not limited solely to Marconigrafia. A survey of nineteenth-century inventors and mothers would surely include the example of a deaf Ma Bell teaching her son, Alexander Graham, to hear; James Clerk Maxwell's mother, who "taught him to read, encouraged (as did his father) his curiosity in all things, and took pride in his wonderful memory"; and, of course, Nikola Tesla, who always said that "he inherited his photographic memory and his inventive genius from his mother." See C. W. F. Everitt, *James Clerk Maxwell: Physicist and Natural Philosopher* (New York: Charles Scribner's Sons, 1975), 41; and Margaret Cheney, *Nikola Tesla: Man Out of Time* (Englewood Cliffs, NJ: Prentice-Hall, 1981), 7. In Oliver Lodge's case, it was a grandmother: "He learned to read very early, probably at the age of three, and his maternal grandmother . . . took particular pleasure in teaching him the art of reading aloud." W. P. Jolly, *Sir Oliver Lodge* (London: Constable, 1974), 12.

40. Landini, *Marconi sulle vie dell'etere,* 29.

41. D. Marconi, *My Father,* 6.

42. Landini, *Marconi sulle vie dell'etere,* 53–54.

43. Alice Yaeger Kaplan, *Reproductions of Banality: Fascism, Literature, and French Intellectual Life* (Minneapolis: University of Minnesota Press, 1986), 136. McLuhan adds a necessary caveat to Kaplan's gloss by noting the wireless feature of all media. "Radio is provided with its cloak of invisibility, like any other medium. It comes to us ostensi-bly with person-to-person directness that is private and intimate, while in urgent fact, it is really a subliminal echo chamber of magical power to touch remote or forgotten chords." McLuhan, *Understanding Media,* 302.

44. Landini, *Marconi sulle vie dell'etere,* 41.

45. Dunlap, *Marconi: The Man and His Wireless,* 38.

46. Landini, *Marconi sulle vie dell'etere,* 41.

47. Ibid., 66.

48. Ibid., 22.

49. Dunlap, *Marconi: The Man and His Wireless,* 3.

50. Ibid.

51. Ibid., 53.

52. Ibid., 4.

53. "In the nearby towns everyone had a big laugh: without a doubt people said that at

Pontecchio a telegraphic epidemic broke out. Everyone is a little nuts" (Landini, *Marconi sulle vie dell'etere,* 15).

54. "Guglielmo led his mother up three flights of shallow, stone steps into his inner world, full of jars and instruments. As she watched, he bowed his blond head over a telegraph key set on a workbench under a window and tapped it delicately with one finger. From the far end of the long double room came a gentle, insistent sound. A bell was ringing, little louder than the crickets but with concise, wakeful clarity. Between the transmitter under his hand and the tiny tinkling lay nothing but air." D. Marconi, *My Father,* 5.

55. Ibid., 36.

56. Ibid., 45–46.

57. Dunlap, *Marconi: The Man and His Wireless,* 46.

58. Ibid.

59. Peter Haining, *The Man Who Was Frankenstein* (London: Frederick Muller Limited, 1979), 5. Preece was for many years an engineer in the Post Office's telegraphic system, helping to devise a railroad signaling system. See Avital Ronell, *The Telephone Book: Technology, Schizophrenia, Electric Speech* (Lincoln: University of Nebraska Press, 1989), 432, n. 6.

60. Ronell, *The Telephone Book,* 365.

61. Solari, *Marconi,* 68. The rezoning method in use here may be more complex than I am proposing. The photograph the young English woman looks at every night suggests itself as another vehicle for spiritual impregnation: the monstrous births associated with photography "emanate from its capacity to catch a resemblance that would not, that could not, be mediated by the lengthy sessions during which both the painter and the model stood waiting for a form of revelation." Marie-Hélène Huet, *Monstrous Imagination* (Cambridge, MA: Harvard University Press, 1993), 214. I am grateful to Barbara Spackman for pointing out the visual-monstrous nexus.

62. Derrida, *The Post Card,* 24.

63. Cf. broadcast: "Of seed, etc.: Scattered abroad over the whole surface instead of being sown in drills or rows; Scattered widely abroad, widely disseminated." *The Compact Edition of the Oxford English Dictionary* (1973), s.v. "broadcast."

64. Solari, *Marconi,* 128–29.

65. Dunlap, *Marconi: The Man and His Wireless,* 21.

66. "He never recovered from his infantile tendency to leave food on his plate. He always had a precarious appetite." D. Marconi, *My Father,* 240.

67. Solari, *Marconi,* 114–15.

68. D. Marconi, *My Father,* 240.

69. Ibid., 256.

70. "From the Hebrew he translates 'tongue,' if you can call it translating, as lip. They wanted to elevate themselves sublimely, in order to impose their lip, the unique lip, on the universe. Babel, the father, giving his name of confusion, multiplied the lips, and this is why we are separated and that right now I am dying, dying to kiss you with our lip the only one I want to hear." Derrida, *The Post Card,* 9.

71. "The problem of detecting such energy is merely one of getting a sufficiently

sensitive device. Who can say that our thoughts are not causing vibrations? Who can set a limit to the distance they may travel, or to the receiving powers of another brain, in some way or other attuned thereto?" Upton Sinclair, *Mental Radio* (New York: Albert & Charles, 1930), 213. And the most sensitive device? "And, as it happened, fate was planning to favor me. It sent me a wife who became interested, and who not merely practiced telepathy, but learned to practice it" (3).

72. "With the telegraph's instantaneous, or 'live' transmissions, the origin of the new media technologies could be attained or retained. At the same time, this kind of 'live' broadcast that the telegraph introduced was the first literalization or realization of telepathy; so it is no coincidence (even though perhaps, after all, it was an accident) that once the telegraph was introduced, we find societywide preoccupation with the occult." Laurence A. Rickels, *The Vampire Lectures* (Minneapolis: University of Minnesota Press, 1999), 53.

73. Dunlap, *Marconi: The Man and His Wireless*, 6.

74. Ibid., 266.

75. Ibid., 327.

76. Ibid., 286. According to a *Vanity Fair* article from 1905, Marconi was "a quiet man with a slow deliberate way of speech and a shape of a head which suggests an unusual brain." Quoted in D. Marconi, *My Father*, insert.

77. D. Marconi, *My Father*, 18.

2. D'Annunzio and the "Marconigram"

1. Gabriele D'Annunzio, *Nocturne and Five Tales of Love and Death*, trans. Raymond Rosenthal (Marlboro, VT: Marlboro Press, 1988), 218. As the English edition contains only selections from D'Annunzio's work, all ensuing translations from the work are my own unless otherwise indicated and are taken from *Notturno* (Milan: Arnoldo Mondadori, 2003).

2. Renato Barilli, *D'Annunzio in prosa* (Milan: Mursia, 1993).

3. Ibid., 8.

4. The offerings in Italian on D'Annunzio's occupation of Fiume are immense. See in particular Guglielmo Ferrero, *Da Fiume a Roma, 1919–23: Storia di quattro anni; L'invenzione del fascismo* (Rome: Nuovi Equilibri, 2003), which links Fiume to fascism in ways typical of Italian historiography; Alessandro Barbero, *Poeta al comando* (Milan: Mondadori, 2002), a recent novel and hagiography of D'Annunzio and the "l'esperienza fiumana"; Claudia Salaris, *Alla festa della rivoluzione: Artisti e liberatori con D'Annunzio a Fiume* (Bologna: Il Mulino, 2002), the most detailed account yet to appear of the social and political aspects of Fiume; Robert Dombroski's chapter on D'Annunzio's mythmaking in *Properties of Writing: Ideological Discourse in Modern Italian Fiction* (Baltimore, MD: Johns Hopkins University Press, 1995); Jeffrey Schnapp, "Le parole del silenzio," *Quaderni D'Annunziani* 3–4 (1988): 35–59; Thomas Harrison, ed., *Nietzsche in Italy* (Saratoga, CA: ANMA, 1988); Barbara Spackman, "*Il verbo (e)sangue*: Gabriele D'Annunzio and the Ritualization of Violence," *Quaderni d'italianistica* 4, no. 2 (1983): 218–29; Umberto Carpi, "Futuristi, metafisici e 'spiriti liberi' nella Fiume di D'Annunzio," *Studi novecenteschi* 8 (December 1981): 133–61; Renzo de Felice, *D'Annunzio politico, 1918–1938* (Rome: Laterza, 1978), the classic account of

the motivations and ideology underlying the entire enterprise; and Ferdinando Gerra, *L'impresa di Fiume* (Milan: Longanesi, 1975), useful for its wealth of photographs and solid archival work. As the reader will soon discover, two accounts in English from al most two decades ago continue to dominate discussions of Fiume: Michael A. Ledeen, *D'Annunzio: The First Duce* (Baltimore, MD: Johns Hopkins University Press, 1977) and George L. Mosse, "The Poet and the Exercise of Political Power: Gabriele D'Annunzio," in *Masses and Man: Nationalist and Fascist Perceptions of Reality* (New York: H. Fertig, 1980), 87–103. A significant contribution to Fiume historiography appeared in German in 1996: Hans Ulrich Gumbrecht, ed., *Der Dichter als Kommandant: D'Annunzio Erobert Fiume* (Munich: Fink, 1996). Of particular interest is Gumbrecht's "I redentori della vittoria: Über Fiumes Ort in der Genealogie des Faschismus" (83–115), which provides a helpful overview of Fiume historiography.

5. Mosse, "The Poet and the Exercise of Political Power," 100.

6. Ledeen, *D'Annunzio*, x. Marc Redfield, reviewing Benedict Anderson's *Imagined Communities,* details the process by which the signs of nationhood interpellate the national subject as the subject of a particular national culture: "The signs of nationhood—the flags and emblems, according to Hegel, in their sheer arbitrariness demonstrate the mind's creative power—serve the cause of misrecognition insofar as they transform a semiotic function (linguistic arbitrariness) into an *image:* an image of the nation as well, or, better, of the nation as imagi-nation" (emphasis in original). "Imagi-nation: The Imagined Community and the Aesthetics of Mourning," *diacritics* 29 (Winter 1999), 66.

7. Ibid., x.

8. Salaris, *Alla Festa della rivoluzione,* 12; Antonio Gibelli, *La grande guerra degli italiani, 1915–1918* (Milan: Sansoni, 1998), 322.

9. De Felice, *D'Annunzio politico, 1918–1938,* 61. See also in this regard N. F. Cimmino, *Poesia ed poetica in Gabriele D'Annunzio* (Florence: Centro internazionale del libro, 1959).

10. Spackman, "Il verbo (e)sangue," 220.

11. Gerra, "L'impresa di Fiume," 500. The speech appears in the indispensable *La penultima ventura: Scritti e discorsi fiumani, a cura di Renzo De Felice,* Gabriele D'Annunzio (Milan: Arnoldo Mondadori Editore, 1974), 354–59. The choice of the name *Elettra* ought not go unremarked as it is one of the most overdetermined in all of Marconigrafia. "'I certainly didn't name the ship *Elettra* to honor Agamemnon's vindictive daughter. *Elettra* from electricity because everything will have to function electrically aboard ship, even the engines,' the Maestro told me with unusual vivacity." Adelmo Landini, *Marconi sulle vie dell'etere: La storica impresa narrata dall'Ufficiale Marconista dell'Elettra* (Turin: Societá Editrice Internazionale, 1955), 108. One might overlook the mythic Elettra if Marconi had not also bestowed the same name on his youngest daughter. If we refuse Marconi's remonstrations in naming the ship for the daughter who avenges the death of her father at the hands of her mother, *Elettra* appears to stage the wireless cutting typical of Marconigrafia that I sketched in the previous chapter. How many false messages, Elettra cries out, must she receive before Oreste returns? Oreste: the message/messenger continually anticipated, and Elettra, the one who waits.

12. Emilio Gentile, "From the Cultural Revolt of the Giolittian Era to the Ideology of

Fascism," in *Studies in Modern Italian History: From the Risorgimento to the Republic,* ed. Frank J. Coppa (New York: Peter Lang, 1986), 112. See too his *Il culto del Littorio: La sacralizzazione della politica nell'Italia fascista* (Rome: Laterza, 1983). Cf. Spackman's notion of symbolic action and violence at Fiume: "The violence which D'Annunzio aestheticizes is not that of Marinetti . . . nor even that of Mussolini's squads or Fascism's later projects. It is instead the violence suffered in WWI, a violence suffered rather than perpetrated which is ritualized in Fiume." Spackman, "Il verbo (e)sangue," 222.

13. In his examinations of Nazism, Saul Friedländer's category of kitsch comes close to what I have in mind: "The important thing is the constant identification of Nazism and death; not real death in its everyday horror and tragic banality, but a ritualized, stylized, and aestheticized death." *Reflections of Nazism: An Essay on Kitsch and Death,* trans. Thomas Weyr (New York: Harper and Row, 1984), 43. Cf. Spackman's rendering of Fiume: "What is striking about the Fiuman writings, which Michael Ledeen has aptly termed 'political passion plays,' is that they are dominated by a Christological rhetoric . . . What is more, it is not a triumphant rhetoric of the Church Militant, but a defeated one of sacrifice and purification." "Il verbo (e)sangue," 220.

14. Ledeen, *D'Annunzio,* 41.

15. G. Rochat, *Gli arditi della grande guerra: Origini, miti, battaglie e miti* (Milan: Feltrinelli, 1981), 130–31.

16. Ferdinando Cordova, *Arditi e legionari dannunziani* (Padua: Marsilio Editori, 1969), 66.

17. Gibelli, *La grande guerra,* 322.

18. Ibid., 323–24.

19. "The *arditi* (crack assault troops, later the backbone of the Fascist squads) are portrayed as being without minds. A confusion of bare arms can be seen pumping like pistons and pouring like taps bottles of wine and glasses of beer amid a fantasmagoric flickering of blue lights and grotesque shadows. Their nonhumanness is conveyed not only by the way they are mechanised, but also by the way in which their surroundings are made unreal, so that the activity of these *arditi* is deprived of any sense or purpose." Christopher Wagstaff, "'Dead Man Erect': F. T. Marinetti, *L'alcova d'acciaio,*" in *The First World War in Fiction,* ed. Holger Klein (London: Macmillan Press, 1976), 155. Cf. Farina: "The *arditi* had to repeat every gesture and every phase of the assault many times until they had acquired a series of automatic reactions, seen not as a guarantee of passive acceptance of destiny, but the conquest of professionalism that could not but have great effects on morale." Salvatore Farina, *Le truppe d'assalto italiano* (Rome: Lavoro fascista, 1938); quoted in Rochat, *Gli arditi della grande guerra,* 37.

20. Gibelli, *La grande guerra,* 138.

21. Niklas Luhmann and Raffaele De Giorgi suggest a similar possibility when discussing medieval heresies. See their *Teoria della società* (Milan: FrancoAngeli, 2000), 98–99. Cf. Zeev Sternhell's perspective on alphabetization and national identity: "Very soon, it appeared that compulsory education, the spread of literacy in the countryside, and the working class's slow but continuous acquisition of culture encouraged not the class consciousness of the proletariat, but rather an increased consciousness of national identity." Zeev Sternhell, *The Birth of Fascist Ideology: From Cultural Rebellion to Political Revolution,* trans. David Maisel (Princeton, NJ: Princeton University Press, 1994), 14.

22. Ledeen, *D'Annunzio*, 39.

23. For a sampling of emerging writers at Fiume, see Sir Robert Oswell, *Nobler Essences: A Book of Characters* (New York: Grosset and Dunlap, 1950), 133–35; Giovanni Comisso, *Le mie stagioni* (Milan: Garzanti, 1951), 49–52. Interestingly, enrolling in the *arditi* entailed a preliminary and explicit written application addressed to their superiors. P. Reginaldo Giuliani, *Gli arditi: Breve storia dei reparti d'assalto della terza armata* (Milan: Fratelli Treves Editore, 1919), 3.

24. D'Annunzio, "Santa Barbara," in *La penultima ventura*, 398.

25. Mosse, "The Poet and the Exercise of Political Power," 96.

26. Elizabeth Eisenstein's comments on the effects of print on consciousness are opportune: "As previous remarks suggest, the effects produced by printing may be plausibly related to an increased incidence of creative acts, to internally transformed speculative traditions, to exchanges between intellectuals and artisans, and indeed to each of the contested factors in current disputes." Elizabeth Eisenstein, *The Printing Press as an Agent of Change: Communications and Cultural Transformations in Early-Modern Europe*, vol. 2 (Cambridge: Cambridge University Press, 1979), 688. In this regard, see also Roddey Reid, *Families in Jeopardy: Regulating the Social Body in France* (Stanford, CA: Stanford University Press, 1993): "The reading of best sellers and the daily consumption of newspapers by the middle class and workers alike amounted to identical rituals of imagining a national community, however internally differentiated or divided it may be" (139); quoted in Jonathan Culler, "Anderson and the Novel," *diacritics* 29 (winter 1999): 26. For the Italian context, see Giorgio Raimondo Cardona, "La linea d'ombra dell'alfabetismo ai confini tra oralità e scrittura," in *Sulla vie della scrittura: Alfabettizzazione, cultura scritta, e istituzioni in età moderna*, ed. Maria Rosaria Pelizzari (Naples: Edizioni Scientifiche Italiane, 1989), 39–54. For an altogether convincing rebuttal of Eisenstein, see Michael Warner, *The Letters of the Republic: Publication and the Public Sphere in Eighteenth-Century America* (Cambridge, MA: Harvard University Press, 1990), 5.

27. De Felice makes the distinction as does Spackman. De Felice, *D'Annunzio politico, 1918–1938*, 27.

28. N. Katherine Hayles, "Simulating Narratives: What Virtual Creatures Can Teach Us," *Critical Inquiry* 10 (April 1999): 10.

29. Mosse, "The Poet and the Exercise of Political Power," 93.

30. Ibid., 92. This is not to say that D'Annunzio's own gifts as an orator were not crucial for generating attention among the recently alphabetized: "D'Annunzio's political effectiveness derived from a variety of elements of his own personality. His love affairs made him a symbol of virility. . . . His use of language and his oratorical skill made him an effective campaigner and an inspirational leader." Ledeen, *D'Annunzio*, 11.

31. D'Annunzio, "Qui rimarremo ottimamente," in *La penultima ventura*, 129.

32. "Listen to me. Yet again there are two cases. Fiume resembles our old trenches. The cases are always two, also here." D'Annunzio, "La sagra di tutte le fiamme," in *La penultima ventura*, 337.

33. D'Annunzio, "All'erta!" in *La penultima ventura*, 262.

34. D'Annunzio, *Notturno*, 228.

35. D'Annunzio, "L'urna inesausta," in *La penultima ventura*, 180.

36. D'Annunzio, "Domando alla città di vita un atto di vita," in *La penultima ventura,* 309.

37. Giorgio Agamben, *Remnants of Auschwitz: The Witness and the Archive,* trans. Daniel Heller-Roazen (New York: Zone Books, 2003). "Thus begins the semantic migration by which the term 'holocaust' in vernacular languages gradually acquires the meaning of the 'supreme sacrifice in the sphere of complete devotion to sacred and superior motives'" (30).

38. Spackman, "Il verbo (e)sangue," 221.

39. D'Annunzio, "Discorso," in *La penultima ventura,* 329; "Agli arditi di Fiume e d'Italia," in ibid., 297.

40. Jacques Derrida, "Of an Apocalyptic Tone Recently Adopted in Philosophy," *Semeia* 23 (1982): 63–98. The Kant essay may be found in Peter David Fenves, *Raising the Tone of Philosophy* (Baltimore, MD: Johns Hopkins University Press, 1993): 51–72.

41. Christopher Norris, "Versions of Apocalypse," in *Apocalypse Theory and the Ends of the World,* ed. Malcolm Bull (Oxford, England: Blackwell, 1995), 240.

42. "Apocalypse," *The Oxford English Reference Dictionary* (Oxford: Oxford University Press, 2002); Derrida, "Of an Apocalyptic Tone," 65.

43. Derrida, "Of an Apocalyptic Tone," 72–73.

44. Ibid.

45. Ibid., 69.

46. Ibid., 83.

47. "Maintaining and joining, the telephone line holds together what it separates. It creates a space of signifying breaks and is tuned by the emergency feminine on the maternal cord reissued." Avital Ronell, *The Telephone Book: Technology, Schizophrenia, Electric Speech* (Lincoln: University of Nebraska Press, 1989), 4.

48. Derrida, "Of an Apocalyptic Tone," 72.

49. Ibid., 69.

50. Ibid., 84. Borrowing from Freud, we might say that the voice of the oracle resembles not so much an obsessional idea as a pathological formation that arises as part of what Freud typically calls "the secondary defensive struggle." The result is the same: reason is delirious. The relevant passage from the "Rat Man" follows: "They [the psychological formations] are not purely reasonable considerations which arise in opposition to the obsessional thoughts, but, as it were, hybrids between the two species of thinking: they accept certain of the premises of the obsession they are combating, and thus, while using the weapons of reason, are established upon a basis of pathological thought. I think such formations as these deserve to be given the name of 'deliria.'" Sigmund Freud, *Three Case Histories* (New York: Touchstone, 1963), 59.

51. Derrida, "Of an Apocalyptic Tone," 84.

52. Christopher Rowland, "'Upon Whom the Ends of the Ages Have Come': Apocalyptic and Interpretation of the New Testament," in *Apocalypse Theory and the Ends of the World,* ed. Malcolm Bull (Oxford, England: Blackwell, 1995), 47.

53. D'Annunzio, "Disobbedisco," in *La penultima ventura,* 87.

54. D'Annunzio, "Agli arditi di Fiume e d'Italia," in *La penultima ventura,* 293.

55. Ibid.

56. Ibid.

57. D'Annunzio, "Proclama," in *La penultima ventura,* 444.

58. D'Annunzio, "La sera dei ribelli," in *La penultima ventura,* 110.

59. D'Annunzio, "Il commiato fra le tombe," in *La penultima ventura,* 461.

60. Schnapp, "Le parole del silenzio," 38.

61. D'Annunzio, "Il commiato fra le tombe," in *La penultima ventura,* 463.

62. The notions of hearing and obeying are linked etymologically, which allows the speeches to traffic in more tautologies: "The languages that relate hearing to the invading features of sound often consider the auditory presence as a type of 'command.' Thus *hearing* and obeying are often united in root terms. The Latin *obaudire* is literally meant as *listening* 'from below.' It stands as a root source of the English *obey.* Sound in its commanding presence *in-vades* our experience" (emphasis in original). Don Ihde, *Listening and Voice: A Phenomenology of Sound* (Athens: Ohio University Press, 1976), 81.

63. Attempting a hazardous and improvised landing on the sea near Grado on January 16, 1916, D'Annunzio suffered a severe blow to the right side of the head that resulted in the loss of the eye. The severity of the accident was not immediately apparent to D'Annunzio, however, as he continued to fly missions and offer orations. See Annamaria Andreoli's introduction to *Diari di Guerra, 1914–1918* (Milan: Oscar Mondadori: 2002), v–xlix.

64. D'Annunzio, *Notturno,* 135.

65. Ibid., 72.

66. Ibid., 137, 70, 31, 7, 18, 52, 5, 58.

67. Ibid., 62.

68. D'Annunzio, "La fiamma intelligente," in *La penultima ventura,* 369.

69. D'Annnuzio, "L'Italia alla colonna e la vittoria col bavaglio," in *La penultima ventura,* 60.

70. D'Annunzio, "Dalla loggetta del Sansovino nel giorno di San Marco," in *La penultima ventura,* 49.

71. Marshall McLuhan, *Understanding Media: The Extensions of Man* (New York: McGraw-Hill, 1967). For Italian commentary, see Remo Ceserani, *Raccontare il postmoderno* (Turin: Bollati Boringhieri, 1998) and his eclectic *Treni di carta: L'immaginario in ferrovia: L'irruzione del treno nella letteratura moderna* (Genoa: Marietti, 1993), as well as Gianni Vattimo, *The Transparent Society,* trans. David Webb (Baltimore, MD: Johns Hopkins University Press, 1992) and *The Ends of Modernity,* trans. Jon R. Snyder (Baltimore, MD: Johns Hopkins University Press, 1991).

72. See "Il mio pomeriggio con Marconi, il mago che incanta le onde," *Corriere della sera,* April 11, 2003, 35. I am grateful to Barbara Valotti for providing me with the details of Marconi and D'Annunzio's first meeting.

73. D'Annunzio, "Saluto a Guglielmo Marconi in Fiume d'Italia," in *La penultima ventura,* 354.

74. Avital Ronell, *Crack Wars: Literature Addiction Mania* (Lincoln: University of Nebraska Press, 1992), 79.

75. D'Annunzio, "Saluto," in *La penultima ventura,* 354.

76. Ibid., 356. An ancient example of mental telepathy, Arpocrate suggests the rapport

the wireless enjoys with unmediated broadcasts to the brain. "It was one of the numerous forms in which the Egyptian god Oro or more precisely the young Oro was venerated. As such, the Egyptian monuments, and later the Greeks and Romans, represented him with a finger in his mouth, which Plutarch interpreted as *a symbol of silence and to be understood in relation to the mysteries* [emphasis mine]. Arpocrate was popular in the last period of Egyptian independence because he spoke to the heart as a symbol of the third constitutive element of the family." *Enciclopedia Cattolico* (1949), s.v. "Arpocrate."

77. For a genealogy of the petrified Marconi operator, Landini's travel log of his Vesuvius expedition is unrivaled. While waiting for Marconi to return from London, the faithful assistant waits again, this time at Pompei. "Yet again: a very ancient city, completely conserved and liberated by the ashes of Vesuvius: Pompei . . . going round the streets of Pompei the sensation is such . . . that one almost expects from one moment to the next to meet up with some former inhabitant. Moreover, one can see all the inhabitants but they are all petrified." Landini, *Marconi sulle vie dell'ettere*, 110.

78. D'Annunzio, "Saluto," in *La penultima ventura*, 355.

79. D'Annunzio, "Domando alla città di vita un atto di vita," in *La penultima ventura*, 312.

80. D'Annunzio, "Saluto," in *La penultima ventura*, 356.

81. Ibid.

82. D'Annunzio, "L'ala d'Italia è liberata," in *La penultima ventura*, 97.

83. D'Annunzio, "Anche questa prova sarà superata," in *La penultima ventura*, 393.

84. D'Annunzio, "Saluto," in *La penultima ventura*, 358.

85. Ibid. Interestingly, Marconi's arteries continually emerge in accounts of wireless as enervated sites, no longer pulsing with life. This forces Marconi into the ranks of the undead. Consider the deathbed conversation between Marconi and his doctor. "Reclining and looking very pale he lifted his forearm and saw that the blood in the artery was not beating any more. Turning to me he said in a low voice, 'How is it, Frugoni, that my heart has stopped beating and I am still alive?' To which I replied: 'Don't ask such questions, it is only a matter of position, because your forearm is raised.' With a little wry smile he said: 'No my dear doctor, this would be correct for the veins but not for an artery,' showing that, to him, a scientist, one could not tell pitiful lies which broke the laws of physics. And in fact he was perfectly correct in his conclusion." Degna Marconi, *My Father, Marconi* (New York: McGraw Hill, 1962), 309.

86. D'Annunzio, "Saluto," *La penultima ventura*, 358.

87. Ibid.

88. Ibid., 359.

89. This conclusion is at odds with J. Hillis Miller's introduction to Derrida. Bent on equating universality with the wireless or scenes of writing generally, Miller confounds destinations with transmissions so that the lack of an address accounts for a transmission's universal characteristic: universality is just another name for nothing left to lose. The wireless transmission at Fiume demonstrates, on the contrary, that if transmissions always refer to other transmissions, an assured address that is to come, then attempts at interpreting transmissions as universal is yet another instance of the oracle deliriously inflecting the voice of reason to its own ends. J. Hillis Miller, "Thomas Hardy, Jacques

Derrida, and the 'Dislocation of Souls,'" in *Taking Chances: Derrida, Psychoanalysis, and Literature,* ed. Joseph H. Smith and William Kerrigan (Baltimore, MD: Johns Hopkins University Press, 1984), 135–45.

90. D'Annunzio, "L'Italia alla colonna e la vittoria col bavaglio," in *La penultima ventura,* 59.

91. No event associated with transmission technologies more clearly summarizes its connection to writing, writing machines, or the message to mother than the translation required by the state for patent approval. The Marconi wireless distinguishes itself in the sheer transparency of the operation. Witness the scene of patent dictation between Anne Jameson and Marconi when typewriters were no match for a mother's longhand and Marconi still had not given up his mother medium. "He was mature enough, however, to undertake the exposition needed for patents with persevering care. 'I had to protect my invention against every possible counterfeit and against a variation of secondary importance.' So he sat and wrote what had to be written in that fluid hand which suggested a race to keep up with the quick processes of his mind, and his mother copied it all in her fine, copperplate script. The typewriter had already been made practicable, but all of the early agreements of my father's which I have seen were laboriously written." D. Marconi, *My Father,* 38.

92. D'Annunzio, "Saluto," in *La penultima ventura,* 357.

93. David E. Wellbery evokes a new posthermeneutic corporeality that corresponds to the kind of bodies the wireless sets about constructing: "The body is not first or foremost an agent or actor, and in order to become one it must suffer a restriction of its possibilities. . . . As a result, culture is no longer viewed as a drama in which actors carry out their various projects. Rather the focus of analysis shifts to the processes that make the drama possible: to the writing of the script, the rehearsals and memorizations, the orders that emanate from the directorial authority." David E. Wellbery, introduction to *Discourse Networks 1800/1900,* by Friedrich Kittler (Stanford, CA: Stanford University Press, 1990), xv.

94. Luigi Solari, *Marconi: Dalla Borgata di Pontecchio a Sydney D'Australia* (Naples: Alberto Morano Editore, 1927), 121.

95. D'Annunzio, "Saluto," in *La penultima ventura,* 355.

96. Or perhaps not. "It sees itself, the response, dictated to be poetic, by being poetic. And for that reason, it is obliged to address itself to someone, singularly to you but as if to the being lost in anonymity, between city and nature, and imparted secret, at once public and private, *absolutely* one and the other." Jacques Derrida, "Che cos'è la poesia?" *The Derrida Reader,* trans. Peggy Kamuf (New York: Columbia University Press, 1991), 289.

97. "I keep thinking about all the kids who got wiped out by seventeen years of war movies before coming to Vietnam to get wiped out for good. You don't know what a media freak is until you've seen the way a few of those grunts would run around during a fight when they knew that there was a television crew nearby; they were actually making war movies in their heads, doing little guts-and-glory Leatherneck tap dances under the fire, getting their pimples shot off for the networks." Michael Herr, *Dispatches* (New York: Vintage International, 1991), 209.

98. Friedrich Kittler, "Media Wars: Trenches, Lightning, Stars," in *Literature Media Information Systems*, ed. John Johnston (Amsterdam: Overseas Publishers Association, 1997), 117. "War, as opposed to sheer fighting has been for a long time an affair of persuasion. It came only into being when people succeeded in making others die for them. Before their probably fatal orders were obeyed, commanders had first to create subjects, both in the philosophical and political sense. Instructions, therefore, presupposed injunctions and these addressed an art of rhetoric."

99. "And why? Because his ears have been opened and he now can hear what is in accord with his nature." Martin Heidegger, *What Is Called Thinking*, trans. J. Glenn Gray and F. Wieck (New York: Harper and Row, 1968), 48; quoted in Ronell, *The Telephone Book*, 27. Although space does not permit a full exposition of the function of the mother's mouth at Fiume, I would be amiss if I did not indicate provisionally the extent to which D'Annunzio's own mother haunts the speeches and *Notturno*. See in particular her ability to speak through him in a bizarre moment of ventriloquy while D'Annunzio stands vigil over yet another mutilated comrade: "My mother speaks through my mouth, speaking to him as his mother speaks to him" (*Notturno*, 137). In the Fiume speeches, Italy is figured as a mother who speaks through D'Annunzio in order to be heard again by her insensitive sons. For the best though by no means only example, see the speech "La sera di ribelli," in *La penultima ventura*, 408–11.

100. See note 92.

101. D'Annunzio, "Saluto," in *La penultima ventura*, 357.

102. The ontically challenged Marconi also doubles for D'Annunzio. "Finding himself therefore in Rome, he decided one evening to visit an important personage as it was growing dark. . . . But a few hundred meters from the house of this very important figure, Marconi's automobile was stopped by two undercover guards who asked him his name. He was authorized to proceed, but after going along a hundred meters, he was stopped again. . . . Using the semidarkness, Marconi said in a grave tone: 'Gabriele D'Annunzio.' 'Stop, stop,' the guards shouted. You may not pass. A brigadier commanding a squad a short distance away was immediately informed of the incident. The brigadier went forward with a decisive and solemn step, but when he reached the car door, Marconi laughed and said: 'But can't you see that it's me, Marconi?' 'Honorable Senator,' responded the brigadier, 'Please don't play such tricks on us. Proceed.'" Solari, *Marconi*, 59–60.

103. Alice Yaeger Kaplan, *Reproductions of Banality: Fascism, Literature, and French Intellectual Life* (Minneapolis: University of Minnesota Press, 1986), 141; quoted in Ronell, *The Telephone Book*, 418.

104. "The present has become suffused with a critical character. It is a moment of utter significance within history and cannot be regarded with detached equanimity. A commitment is necessary, and action must follow from commitment to the cause. It is a moment pregnant with opportunity for fulfilling the destiny of humankind. Here indeed is the moment when heaven finally comes on earth." Christopher Rowland, "Upon Whom the Ends of the Ages Have Come," 42.

105. Ronell, *The Telephone Book*, 412.

106. In what is perhaps the only indisputable point in his recent essay on fascism,

Umberto Eco imagines the fate of D'Annunzio under the Nazis or Bolsheviks: "The Italian national poet was D'Annunzio, a fop who in Germany or Russia would have found himself in front of a firing squad. He was elevated to the rank of bard to the regime for his nationalism and cult of heroism—with the addition of a strong dash of French decadence." Umberto Eco, "Ur-fascism," in *Five Moral Pieces*, trans. Alastair McEwen (London: Vintage, 2002), 74.

107. Renzo de Felice, "Introduzione," vii. The most he will admit is that Fiume stimulated psychological and moral reactions through its "very particular atmosphere," which serves again to distance the occupation from fascism (de Felice, *D'Annunzio politico, 1918–1938*, 35). Gumbrecht adopts much the same perspective in *D'Annunzio Erobert Fiume*.

108. Derrida, "Of an Apocalyptic Tone," 72.

109. Only a handful of studies examine Mussolini or more generally fascism's use of apocalyptic imagery and rhetoric and then only indirectly. Cf. the analyses of Zeev Sternhell and George L. Mosse. For the former, fascism incarnates Sorel's great insight: "[T]he masses need myths in order to go forward. It is sentiments, images, and symbols that hurl individuals into action, not reasonings. It was likewise from Sorel in particular and the Sorelians in general that fascism borrowed something else: the idea that violence gave rise to the sublime" (Sternhell, *The Birth of Fascist Ideology,* 28). Mosse is closer to my own position here when discussing temporality and apocalypse in Nazi Germany: "We seem to be back with the telescoping of history so familiar in earlier ages and through the Renaissance. But now such telescoping fastens onto an apocalyptic tradition and is a conscious attempt to abolish the flow of time that led to the dilemmas of modernity." George L. Mosse, "Death, Time, and History," in *Masses and Man,* 85.

110. For my use of "program," see Derrida's discussion of the Nazi orchestration of the Nietzschean opus in *The Ear of the Other: Otiobiography, Transference, Translation,* ed. Christie McDonald, trans. Peggy Kamuf (Lincoln: University of Nebraska Press, 1985). "The question that poses itself for us might take this form: Must there not be some powerful utterance-producing machine that programs the movements of the two opposing forces at once, and which couples, conjugates or marries them in a given set, as life (does) death?" (29).

111. As early as 1917, "Mussolini came to the conclusion that the defense of the country, the promotion of its influence, and the requirements of a revolution in the true sense (that is, one like the French Revolution or the first stage of the Russian Revolution, which was patriotic and took up the defense of the nation), necessitated a dictatorhsip" (Sternhell, *The Birth of Fascist Ideology,* 220).

112. A sampling would surely include the following speeches in *Mussolini as Revealed in His Political Speeches,* trans. Barone Bernardo Quaranta di San Severino (New York: E. P. Dutton, 1923): "Fascismo's Interests for the Working Classes" (75–81) ["They are dead; they have fallen. But we, in this great hour of your history, O people of Ferrara, will recall them one by one in the orders of the day; and since they are not dead, because their immortal clay is transformed in the infinite play of the possibilities of the universe, we ask of the pure, bright blood of the youth of Ferrara the inspiration to be true to our ideals, to be faithful to the nation" (78)]; "The Government of Speed" (234);

"Men Pass Away, Maybe Governments Too, But Italy Lives and Will Never Die" (323–25) ["Thousands and thousands of those who suffered martyrdom in the trenches, who have resumed the struggle after the war was over, who have won—all those have ploughed a furrow between the Italy of yesterday, of to-day and of to-morrow"]. For later technologically sanctioned prophecy, see "Ritornate," in *Opera Omnia di Benito Mussolini,* vol. 32, *Dalla liberzione di Mussolini all'epilogo la Repubblica Sociale,* ed. Edoardo and Duilio Susmel (Florence: La Fenice, 1960), 87–92; and "A quattrocento ufficiali della guardia nazionale repubblicana," in ibid., 162–66.

113. Ledeen, *D'Annunzio,* 201.

114. Gentile, "From the Cultural Revolt of the Giolittian Era to the Ideology of Fascism," 114–15.

115. Ledeen, *D'Annunzio,* 202; Mosse, "The Poet and the Exercise of Political Power," 93; de Felice, *D'Annunzio politico, 1918–1938,* 139.

116. Gentile, "From the Cultural Revolt of the Giolittian Era to the Ideology of Fascism," 115.

3. STATE OF THE ART

1. Agostino Gemelli, "L'esperimento in psicologia: Del suo valore e dei suoi limiti," *Rivista di psicologia applicata* 4 (1908): 53, 62, 64.

2. "Blindness and deafness, precisely when they affect either speech or writing, yield what would otherwise be beyond reach: information on the human information machine." Friedrich A. Kittler, *Gramophone, Film, Typewriter,* trans. Geoffrey Winthrop-Young and Michael Wutz (Stanford, CA: Stanford University Press, 1999), 189.

3. F. T. Marinetti, "The Technical Manifesto of Futurist Literature," in *Marinetti: Selected Writings,* trans. R. W. Flint and Arthur A. Coppotelli (New York: Farrar, Straus and Giroux, 1971), 84–89; Jeffrey T. Schnapp, "Propeller Talk," *Modernism/Modernity* 1 (1993): 153–78.

4. "The irregular oscillatory movement observed in microscopic particles or 'molecules' of all kinds." *The Compact Edition of the Oxford English Dictionary* (1971), s.v. "Brownian."

5. Jeffrey T. Schnapp, "Crash (Speed as Engine of Individuation)," *Modernism/ Modernity* 6 (1999): 3. For recent work on the cultural history of speed, see Schnapp's previously cited "Propeller Talk," to which I am particularly indebted; Hal Foster's "Prosthetic Gods," *Modernism/Modernity* 4 (1997): 5–38, for a study of how Marinetti and Wyndham Lewis convert the shock of speed into a subject that *thrives* on speed; and Frank Helleman's analysis of the speed of the wireless, "Towards Techno-Poetics and Beyond: The Emergence of Modernism/Avant-Garde Poetics out of Science and Media-Technology," in *The Turn of the Century: Modernism and Modernity in Literature and the Arts,* European Cultures: Studies in Literature and the Arts, no. 3, ed. Christian Berg and Frank Durieux (Berlin: de Gruyter, 1995), 291–301. Although dated in some respects, Stephen Kern's chapter on speed is still the best introduction to how the "barrage of speed brought out the dark side of modernity" (124). Stephen Kern, *The Culture of Time and Space* (Cambridge, MA: Harvard University Press, 1983), 109–30.

6. These include William Preyer, Hermann Helmholtz, and the great psycho-

physicist Fechner. For a synopsis of their work, Ernst Jünger's classic essay "Drugs and Ecstasy" remains the best introduction. In *Myths and Symbols: Studies in Honor of Mircea Eliade,* ed. Joseph M. Kitagawa and Charles H. Long (Chicago: University of Chicago Press, 1962), 327–42.

7. Giuseppe Sergi, *Teoria fisiologica della percezione: introduzione allo studio della psicologia* (Milan: Fratelli Dumolard, 1881), 225.

8. Ibid, 226.

9. Ibid.

10. Ibid., 227.

11. "Something more tangible is at stake: the fact that the readability of signs is a function of their spatiality. The architect's manipulation of space demonstrates that, when the lack is lacking and no empty spaces remain, media disappear, 'naked and obscene,' into the chaos from which they are derived." Friedrich Kittler, *Discourse Networks, 1800/1900,* trans. Michael Metteer (Stanford, CA: Stanford University Press, 1990), 257.

12. Sergi, *Teoria fisiologica della percezione,* 229. Sergi's physiology of the ear's reproduction is deeply indebted to the work of William Preyer. See Preyer's *Über die Grenzen der Tonwahrnehmung* (Jena, Germany: H. Dufft, 1876); and *The Mind of the Child: Observations concerning the Mental Development of the Human Being in the First Years of Life,* trans. H. W. Brown (New York: Appleton, 1888–89).

13. Wellbery, foreword to Kittler, *Discourse Networks,* xv.

14. Sergi, *Teoria fisiologica della percezione,* 229.

15. "Code should be well spaced, so that the receiver is required to make no effort in its translation. Where the receiver is forced to solve a puzzle in a way of improper or badly used code, the context of the story is lost and a break usually follows." Walter P. Phillips, *The Phillips Code: A Thoroughly Tested Method of Shorthand Arranged for Telegraphic Purposes and Contemplating the Rapid Transmission of Press Reports; Also Intended to Be Used as an Easily Acquired Method for General Newspaper and Court Reporting* (New York: Telegraph and Telephone Age, 1907, rev. 1923), unnumbered.

16. Marinetti, "Technical Manifesto," 84. Cf. Russolo's "The Art of Noises": "Let us cross a great modern capital with our ears more alert than our eyes, and we will get enjoyment from distinguishing the eddying of water, air and gas in metal pipes, the grumbling of noises that breathe and pulse with indisputable animality, the palpitation of valves, the coming and going of pistons, the howl of mechanical saws, the jolting of a tram on its rails, the cracking of whips, the flapping of curtains and flags." Luigi Russolo, "The Art of Noises (extracts) 1913," in *Futurist Manifestos,* ed. Umbro Apollonio, trans. R. W. Flint, J. C. Higgitt, and Caroline Tisdall (New York: Viking Press, 1970), 85.

17. Marinetti, "Technical Manifesto," 88.

18. My argument owes much to Jeffrey Schnapp's reading of the propeller, aviation, and the way in which the "Technical Manifesto" gives expression to the cognitive possibilities opened by the invention of the propeller. "Dictation taken from an object. A cutting-edge object, to be sure, but a mere object, all the same, and one that is infringing on the place usually reserved for Marinetti's blunt-edged egocentric muse. Yet the manifesto is unambiguous: a propeller, not a poet, will do the talking; a propeller will

dictate from on high the laws that are to govern modern poetic discourse and define the desublimated forms of individuality and subjectivity attached thereto." Jeffrey T. Schnapp, "Propeller Talk," 153–54.

19. Marinetti, "Technical Manifesto," 87.

20. *Dizionario etimologico della lingua italiana* (1988), s.v. "Sorprendere."

21. Marinetti, "Technical Manifesto," 87. For my use of the term "information," see Weaver: "The concept of information applies not to the individual messages (as the concept of meaning would), but rather to the situation as a whole, the unit information indicating that in this situation one has an amount of freedom of choice, in selecting a message, which it is convenient to regard as a standard or unit amount." Warren Weaver, "Some Recent Contributions to the Mathematical Theory of Communication," in *The Mathematical Theory of Communication,* by Claude E. Shannon (Chicago: University of Illinois Press, 1963), 9.

22. Marinetti, "Technical Manifesto," 84.

23. On Marinetti's penchant for marketing, see Claudia Salaris and Lawrence Rainey, "Marketing Modernism: Marinetti as Publisher," *Modernism/Modernity* 1 (September 1994): 109–27.

24. Marinetti, "Technical Manifesto," 89; emphasis in original. The English translation of *l'immaginazione senza fili* as "the imagination without strings" is unfortunate as it obscures the term's invocation of Marconi's *telegrafia senza fili.* For this reason, I have chosen throughout to refer to the term as "wireless imagination."

25. Ibid., 85; emphasis in original.

26. Marinetti, "Risposte alle obiezioni," in *Teoria e invenzione futurista,* ed. Luciano De Maria (Milan: Mondadori, 1968), 56. Arguably more significant than even the "Technical Manifesto" itself, this short rejoinder has yet to be translated into English. All translations of the "Risposte" are thus my own.

27. Kittler, *Discourse Networks,* 113. Cf. A. A. Mendilow's distinction between Classicism and the Romantics: "The Romantics, on the other hand, saw significance rather in the creative temper that went to the forming of one state to another." *Time and the Novel* (New York: Humanities Press, 1972), 4.

28. Marinetti, "Technical Manifesto," 86; emphasis in original.

29. See Martin Jay, *Downcast Eyes: The Denigration of Vision in Twentieth-Century French Thought* (Berkeley: University of California Press, 1993). For a survey of Italian commentary, compare Gianfranco Contini, *Letteratura dell'Italia Unita, 1861–1968* (Florence: Sansoni, 1968), 666–68; as well as "Innovazioni Metriche Italiane fra Otto e Novecento," in *La letteratura italiana Otto-Novecento,* vol. 4 (Florence: Sansoni, 1974), 185–95; Pier Vicenzo Mengaldo, *Il Novecento* (Milan: Il Mulino, 1991), 206–10; Walter Binni, *La poetica del decadentismo italiano* (Florence: Sansoni, 1938); Carlo Bo, "La nuova poesia," in *Il Novecento,* ed. Natalino Sapegno (Milan: Garzanti, 1987). A useful summary of Futurist commentators may be found in *Dizionario critico della letteratura italiana,* 2d ed., vol. 2 (Turin: Unione Tipografico Editrice, 1986), s.v. "Futurismo."

30. See the 1914 manifesto, "Geometric and Mechanical Splendor and the Numerical Sensibility," for the distinction between the two: "The words-in-freedom, in this continuous effort to express with the greatest force and profundity, by means of free, expressive

orthography and typography, the synoptic tables of lyric values and designed analogies." F. T. Marinetti, "Geometric and Mechanical Splendor and the Numerical Sensibility," in *Marinetti: Selected Writings*, 100. As John J. White points out, however, Marinetti's depiction may not be quite accurate. John J. White, "The Argument for a Semiotic Approach to Shaped Writing: The Case of Italian Futurist Typography," *Visible Language* 10 (Winter 1976): 83.

31. Jeffrey T. Schnapp offers a reading of visual perception in the manifesto, framed by the impact of aviation on the collective imagination: "It is under the aegis of this mechanized human type . . . that I now wish to turn away from the 1912 manifesto's musings on the transfiguration of matter towards what was earlier defined as their visual counterpart, a *poetics of image-streams* [emphasis mine] associated with aviation's impact on the faculty of sight." Schnapp, "Propeller Talk," 165–66. Schnapp goes on to examine "networks of perceptual and/or intuitive strings" envisaged by flight plans and wireless telegrams but never makes explicit the nature of the wireless medium that successfully picks up the propeller's transmission.

32. F. T. Marinetti, "Destruction of Syntax—Wireless Imagination—Words-in-Freedom," in *Futurist Manifestos,* trans. Robert Brain, R. W. Flint, et al. (New York: Viking Press, 1970), 98.

33. While Schnapp is surely correct that "in Futurism's subsequent history, shocks, or, as they were often referred to in the scientific literature, *thrills* . . . are actively sought out: sought out on the highways; in the trenches; in galleries, theaters and stadiums, in the streets" (Schnapp, "Crash," 6; emphasis in original), for the Futurist, the wireless is the medium that ensures that these thrills or shocks are registered and written down.

34. This is natural given speed's etymological roots in the Old English *sped,* which is generally defined as "success or prosperity," and the Old Saxon *spodian,* "to cause to succeed." *A Comprehensive Etymological Dictionary of the English Language: Dealing with the Origin of Words and Their Sense Development Thus Illustrating the History of Civilization and Culture* (1966), s.v. "Speed."

35. Carlo Belloli described Marinetti's method in *Le soir* as simply the attempt "to exhaust all the possibilities that typography has of reproducing an experience, in order to open up new paths for it in the future." The same may be argued for Futurist aesthetic strategies generally; the wireless command is to register exhaustive experience. "Pionere der Grafik in Italien/Italian Pioneers of Graphic Design/ Pionniers du Graphisme en Italie," *Neue Grafik—New Graphic Design—Graphisme Actuel* 3 (October 1959): 9.

36. Friedrich Kittler, "Media Wars: Trenches, Lightning, Stars," in *Literature Media Information Systems,* ed. John Johnston (Amsterdam: Overseas Publishers Association, 1997), 102.

37. Marinetti, "Destruction of Syntax," in *Futurist Manifestos,* 98.

38. Ibid., 101; emphasis in original. The war in Libya served not only to install wireless imaginations in former poets but was also a general testing site for future cinematic and military techniques, especially the trolley *(carrello)* shot and the use of artificial lighting in outdoor shoots. The former conferred "on the cine-camera autonomy and ductility, rescuing it from the bondage that had condemned it to immobility." Giovanna Finocchiaro Chimirri, *D'Annunzio e il cinema cabiria* (Catania: CUECM, 1986), 45. The

resulting shots of remarkable depth first used in Pastroni's film *Cabiria* are the other side of the military tracking shot that utilized a newly found mobility of the camera to take aim. See Paul Virilio, *War and Cinema: The Logistics of Perception,* trans. P. Camiler et al. (New York: Verso Books, 1997).

39. Marinetti to Palezzeschi, January 1912, *F. T. Marinetti–Aldo Palezzeschi: Carteggio,* ed. Paolo Prestigiacomo (Milan: Arnoldo Mondadori, 1978), 61; quoted in Schnapp, "Propeller Talk," 171. See also Chris Bongie, "Declining Futurism: La Battaglia di Tripoli and Its Place in the 'Manifesto tecnico della letteratura futurista,'" *Quaderni d'Italianistica: Official Journal of the Canadian Society for Italian Studies* 15 (Spring–Autumn 1994): 217–25.

40. Robert Lincoln O'Brien, "Machinery and English Style," *Atlantic Monthly* 94 (1904): 464–72.

41. The curious reader will find many in Paul Fussell's anthology of modern war writings. Fussell brings together both the canonical war poets of the First World War and the swollen numbers of war correspondents from World War II and Vietnam. Paul Fussell, *The Norton Book of Modern War* (New York: W. W. Norton, 1991).

42. Michael Herr, *Dispatches* (New York: Vintage Books, 1991), 22.

43. Ibid., 65.

44. Ibid.

45. Marinetti, "Destruction of Syntax," 99.

46. Hal Foster writes of the double logic governing the machinic imaginary of high modernism that "subsumed in new technologies" while dismembering and reducing bodies (Foster, "Prosthetic Gods," 5). This may account for the absence of the wireless (and Marconi) in a work as exhaustive and encyclopedic as Roberto Tessari's classic reading of myth and utopia in industrializing Italy, *Il mito della macchina: Letteratura e industria nel primo Novecento italiano* (Milan: U. Mursia, 1973), 209. For an updated reading of technology and myth in Marinetti, see Cinzia Sartini Blum, "Transformations in the Futurist Technological Mythopoeia," *Philological Quarterly* 74 (Winter 1995): 77–97.

47. In reaching this conclusion, I am indebted to Alexander J. Field's brilliant analysis of pre-wireless signaling devices, particularly his understanding of the basic transmission set and its relation to the Napoleonic system of optical telegraphy. "Communication over the French system sacrificed some complexity of discourse in order to increase the information content of each signal transmitted and coupled this with a large and complex basic transmission set in order to maximize channel capacity given the inherently high composition and reamplification lags. . . . One of the advantages of his repeaters was that they embodied visual recognition technology giving him the ability to distinguish cheaply among a larger transmission set." Alexander J. Field, "French Optical Telegraphy, 1793–1855," *Technology and Culture* 35 (April 1994): 330–31.

48. Kittler, "Media and Drugs," 110.

49. Marinetti, "Risposte alle obiezioni," 56.

50. See Lance Schachterle's recent article on Pynchon for a succinct history of the term and to whom I am indebted for the following section. "Information Entropy in Pynchon's Fiction," *Configurations* 4 (1996): 189.

51. Ibid., 190. Shannon's classic work is *The Mathematical Theory of Communication* (1949; Urbana: University of Illinois Press, 1998).

52. Jacques Lacan, "The Circuit," in *The Seminar of Jacques Lacan, vol. 2, The Ego in Freud's Theory and in the Technique of Psychoanalysis, 1954–1955* (New York: W. W. Norton, 1991), 83.

53. Marinetti, "Multiplied Man and the Reign of the Machine," in *Marinetti: Selected Writings,* 91.

54. At this point, it only seems fair to recall another of McLuhan's insights into the workings of media, which the wireless and its inflected writing continue to confirm: "For each of the media is also a powerful weapon with which to clobber other media and other groups. The result is that the present age has been one of multiple civil wars that are not limited to the world of art and entertainment." Marshall McLuhan, *Understanding Media: The Extensions of Man* (New York: McGraw-Hill, 1967), 20–21.

55. Lacan, "The Circuit," 83.

56. One of the best recent histories of Maxwell's demon is Bruce Clarke's examination of Maxwell and allegory in "Allegories of Victorian Thermodynamics," *Configurations* 4 (1996): 67–90. See also Martin Goldman, *The Demon in the Aether: The Story of James Clerk Maxwell* (Edinburgh: Paul Harris, 1983); N. Katherine Hayles, *Chaos Bound: Orderly Disorder in Contemporary Literature and Science* (Ithaca, NY: Cornell University Press, 1990), esp. ch. 2; and George Levine, *Realism and Representation: Essays on the Problem of Realism in Relation to Science, Literature, and Culture* (Madison: University of Wisconsin Press, 1993).

57. Weaver, "Some Recent Contributions to the Mathematical Theory of Communication," 8.

58. Hal Foster is surely correct when he notes that "the desire to embrace technology, to accelerate its transformation of bodies and psyches, is not bound to any one cultural politic," for indeed both Antonio Gramsci and Walter Benjamin advocated such an approach (Foster, "Prosthetic Gods," 7). But Foster's elision of the new machines of speed and representation under the rubric "technology" is problematic for it fails to account for the radical nature of the wireless transmission and its primacy in Futurist aesthetic considerations.

59. Cf. Paul Virilio on how the city was founded: "In ancient warfare, defense was not speeding up but slowing down. The preparation for war was the wall, the rampart, the fortress. And it was the fortress as permanent fortification that settled the city into permanence. Urban sedentariness is thus linked to the durability of the obstacle." Following Virilio's lead, we might say that the wireless battlefield will be one linked to the mobility of the former obstacle. Paul Virilio, *Pure War,* trans. Mark Polizzotti (New York: Semiotext(e), 1983), 12.

60. "It is a well-known fact that when a living organism is penalized, by comparison with other representatives of its species, through losing the use of a particular organ or faculty, it is apt to respond to this challenge by specializing in the use of some other organ or faculty until it has secured an advantage over its fellows in this second field of activity to offset its handicap in the first." Arnold J. Toynbee, *A Study of History,* vol. 1 (London: Oxford University Press, 1934), 209. Toynbee offers the example of the lame barbarian

who cannot fight but whose hands can forge weapons and armor for other men to wield and wear. See, in this regard, Hermann Goering, who became a pilot in World War I only after discovering that his rheumatism prevented long infantry marches. Not coincidentally, Marconi also suffered from arthritis and rheumatism. He believed his wireless invention, with its pulsing electromagnetic waves, would not only relieve the aches and pains associated with swollen joints but also help enlist Italians incapable of moving on foot as communication specialists. See the middle section of Landini's biography of Marconi for an extended soliloquy on his turn as a *marconista* assigned to an Italian infantry division. Theirs were not isolated incidents. "While in France the handicapped are exempt from military service, the German army had few or no exemptions, for it had decided to *make physical handicaps functional* by using each man according to his specific disability: the deaf will serve in heavy artillery, hunchbacks in the automobile corps, etc. Paradoxically, the dictatorship of movement exerted on the masses by the military powers led to the promotion of unable bodies." Paul Virilio, *Speed and Politics: An Essay on Dromology,* trans. Mark Polizzotti (New York: Semiotext(e), 1986), 61.

61. Marinetti, "Destruction of Syntax," 106.

62. Ibid.; emphasis in original.

63. Marinetti, "Technical Manifesto," 85; emphasis in original.

64. Temple Grandin, *Thinking in Pictures and Other Reports from My Life with Autism,* with a foreword by Oliver Sacks (New York: Vintage Books, 1995), 31.

65. Marinetti, "Technical Manifesto," 85.

66. Grandin, *Thinking in Pictures,* 29.

67. *Dizionario etimologico della lingua italiana* (1988), s.v. "Fondere." The etymology is from the Latin *fundere, versare,* to pour.

68. Marinetti, "Technical Manifesto," 85.

69. Grandin, *Thinking in Pictures,* 32.

70. Compare Grandin's account with Maupassant's "The Horla": "Was it not possible that one of the imperceptible keys of the cerebral finger-board had been paralyzed in me? Some men lose the recollection of proper names, or of verbs, or of numbers, or merely of dates, in consequence of an accident. The localization of all the avenues of thought has been accomplished nowadays; what, then, would there be surprising in the fact that my faculty of controlling the unreality of certain hallucinations should be destroyed for the time being?" Guy de Maupassant, "The Horla," in *Short Stories of the Tragedy and Comedy of Life* (Akron, OH: St. Dunstan Society, 1903), 22.

71. Marinetti, "Manifesto tecnico della letteratura," in *Teoria e invenzione futurista,* 53. I cite the Italian here because the Flint translation is incomplete. Indeed, two entire paragraphs are excised for reasons unclear and unannounced, proof again of the pressing need for a new translation of the manifestos.

72. Marinetti, "Destruction of Syntax," 98; emphasis in original. Again, the translation is misleading as it misses the clear association with wires and wireless (*fili* in the original).

73. This is perhaps what Jeffrey Schnapp has in mind when he recounts the fascist mass spectacle of Alessandro Blasetti's *18 BL:* "Perhaps, most important of all, in a spectacle within which a few individuals would speak for the nation as a whole, it permitted

amplification. . . . A vocal gigantism could be achieved that would grant the occasional dialogues exchanged among the human protagonists priority over the sea of machine noises." Jeffrey T. Schnapp, *Staging Fascism: 18 BL and the Theater of the Masses for the Masses* (Stanford, CA: Stanford University Press, 1996), 62.

74. Marinetti, "Destruction of Syntax," 97.

75. Ibid.

76. See the shade of Teiresias's speech in Homer, *The Odyssey,* trans. Rodney Merrill (Ann Arbor: University of Michigan Press, 2002), 389.

77. And so unlike the hauntings that followed the wireless. See Donna Haraway's useful distinction between organism and machine, which holds true for the wireless and its various offspring: "Pre-cybernetic machines could be haunted; there was always the spectre of the ghost in the machine. This dualism structured the dialogue between materialism and idealism that was settled by a dialectical progeny, called spirit of history, according to taste." "A Cyborg Manifesto: Science, Technology, and Socialist-Feminism in the Late Twentieth Century," in *Simians, Cyborgs, and Women: The Reinvention of Nature* (New York: Routledge, 1991), 152; quoted in Clarke, "Allegories of Victorian Thermodynamics," 90.

78. What is true of the wireless is true for Futurism: both begin to coincide with war. See, for example, cyclist F. T. Marinetti's undated *Letteratura futurista del volontario ciclista*. With Austrian shells raining down, Marinetti finally finds "absolute Futurism" after spending "8 days under fire wonderfully tortured Life with penetrating cold" (F. T. Marinetti, undated, Museo Civico del Risorgimento, Bologna). Many other examples are available to demonstrate what Paul Virilio has unflinchingly argued for thirty years: "Vain attempts have been made to fit Marinetti's works into various artistic and political categories: but Futurism in fact comes from a single art—that of war and its essence, speed." Virilio, *Speed and Politics,* 62.

4. POUND'S "RADIO *CANTOS*"

1. See in particular points 4 and 5 under the heading "*La radia* will be": "4. The reception amplification and transfiguration of vibrations emitted by living beings or dead spirits dramas of wordless noise-states. 5. The reception amplification and transfiguration of vibrations emitted by matter. Just as today we listen to the song of the forest and the sea so tomorrow shall we be seduced by the vibrations of a diamond or a flower." F. T. Marinetti, "La radia," in *Wireless Imagination: Sound, Radio, and the Avant-Garde,* ed. Douglas Kahn and Gregory Whitehead, trans. Stephen Sartarelli (Cambridge: MIT Press, 1992), 267.

2. The last fifteen years of Poundian studies have seen an explosion of titles dedicated to Pound's relationship to media and fascism. Space limits a full accounting, but the interested reader is directed to the following titles: Margaret Fisher, *Ezra Pound's Radio Operas: The BBC Experiments, 1931–1933* (Cambridge: MIT Press, 2002); Paul Morrison, *The Poetics of Fascism: Ezra Pound, T. S. Eliot, Paul de Man* (New York: Oxford University Press, 1996); John Whittier-Ferguson, *Framing Pieces: Designs of Gloss in Joyce, Woolf, and Pound* (New York: Oxford University Press, 1996) (Morrison and Whittier-Ferguson are both available online from Oxford University Press); Daniel Tiffany,

Radiocorpse: Imagism and the Cryptaesthetic of Ezra Pound (Cambridge, MA: Harvard University Press, 1995); Vincent Sherry, *Ezra Pound, Wyndham Lewis and Radical Modernism* (New York: Oxford University Press, 1993); Jacqueline Kaye, ed., *Ezra Pound and America* (Houndmills, England: Macmillan, 1992), especially L. S. C. Bristow's "God, My God, You FOLKS ARE DUMB!!!: Pound's Rome Radio Broadcasts," 18–42; Michael North, *The Political Aesthetic of Yeats, Eliot, and Pound* (Cambridge: Cambridge University Press, 1991); Tim Redman, *Ezra Pound and Italian Fascism* (Cambridge: Cambridge University Press, 1991); Robert Casillo, *The Genealogy of Demons: Anti-Semitism, Fascism, and the Myths of Ezra Pound* (Evanston, IL: Northwestern University Press, 1988); Bruce Fogelman, *Shapes of Power: The Development of Ezra Pound's Poetic Sequences* (Ann Arbor, MI: UMI Research Press, 1988); Kathryne V. Lindberg, *Reading Pound Reading: Modernism after Nietzsche* (New York: Oxford University Press, 1987); and Peter Nicholls, *Ezra Pound: Politics, Economics, and Unity: A Study of the Cantos* (London: Macmillan, 1984).

3. "One evening we went to see a Ginger Rogers–Fred Astaire film and came home late. All the way home from the cinema Babbo tapped and leapt and encouraged me to do likewise and 'get nimble.' Mamile laughed and we were very gay. As we started to undress we heard a loud fracas in Babbo's room—now that he had thrown off his coat and jacket he leapt and tap-danced more freely. . . . It was hard for him to keep still having fully danced out the rhythm he had absorbed." Mary de Rachewiltz, *Discretions* (Boston: Little, Brown, 1971), 127.

4. D. D. Paige, ed., *The Letters of Ezra Pound, 1907–1941* (London: Faber and Faber, 1951), 442.

5. Ezra Pound, Rapallo, to Homer Pound, November 29, 1924, Paige Collection, Yale University, quoted in Carroll Franklin Terrell, *A Companion Guide to the Cantos of Ezra Pound* (Berkeley: University of California Press, 1980), 75.

6. Pound, "Appunti I. Lettera al Traduttore," *L'indice,* no. 8 (October 1930): 1.

7. David Kahn, *The Codebreakers: The Story of Secret Writing* (London: Weidenfeld and Nicholson, 1967), 298. The student of the wireless and its impact on this century's wars will find no better guide than Kahn's history. The following pages owe much to his work, as they do to O. W. Riegel's indispensable, early history of warfare and radio, *Mobilizing for Chaos: The Story of the New Propaganda* (New Haven, CT: Yale University Press, 1934); and Don E. Gordon's later history, *Electronic Warfare: Element of Strategy and Multiplier of Combat Power* (New York: Pergamon Press, 1981). For an essential overview of technological gains in the interwar period, see Williamson Murray and Allan R. Millett, eds., *Military Innovation in the Interwar Period* (Cambridge: Cambridge University Press, 1996).

8. A postwar commission of enquiry into the Caporetto disaster fixed the blame squarely on intercepted wireless transmissions: "The enemy had known and deciphered all our codes, even the most difficult and the most secret"; quoted in Kahn, *The Codebreakers,* 320. Not only the Italians suffered the effects of wireless interception. Twice during World War I the British had "the exceptional good fortune to gain special insight into the German cryptographic process. . . . During World War I, the British were provided by the Russians with a captured code book from the German cruiser *Magdeburg.*" Gordon, *Electronic Warfare,* 27.

9. Cf. the British Government Code and Cipher School (GCCS) and its operations in tracking German units in World War II. "The GCCS listened to messages between German units whose mobility made them dependent on radio rather than land lines. These messages, it was apparent, were sent in ever-changing gobbledygook over stations transmitting at low strength. So the traffic had to be picked up by extremely sensitive 'ears,' transposed from cipher to plain German, then translated and analyzed." William Stevenson, *A Man Called Intrepid: The Secret War* (New York: Harcourt Brace Jovanovich, 1976), 24.

10. The roles were reversed on the Russian front: "As Russian armies moved farther from their base in August 1914, they were reduced to sending uncoded wireless messages because their staffs lacked codes and cryptographers. While maneuvering near Tannenberg, the Germans began to intercept those messages, and General Ludendorff was emboldened to attack one section of the Russian army after he learned that it could be surrounded and taken without danger of interference from the rest of the Russian army in that sector." Stephen Kern, *The Culture of Time and Space* (Cambridge, MA: Harvard University Press, 1983), 309–10.

11. Kahn, *The Codebreakers,* 300.

12. The lack of an address poses the question of death. "Indeed, a radio-talk is produced in accordance with a very specific modality of speech, since it is addressed to a mass of invisible listeners by an invisible speaker. It may be said that, in the imagination of the speaker, it isn't necessarily addressed to those who listen to it, but equally to everyone, the living and the dead." Jacques Lacan, *The Seminar of Jacques Lacan,* vol. 1, *Freud's Papers on Technique, 1953–1954,* ed. Jacques Alain-Miller, trans. John Forrester (New York: W. W. Norton, 1991), 31. See also Jeffrey Sconce's thematic reading of the wireless and death, *Haunted Media: Electronic Presence from Telegraphy to Television* (Durham, NC: Duke University Press, 2000); and Lawrence Rainey, "Taking Dictation: Collage Poetics, Pathology, and Politics," *Modernism/Modernity* 5 (1998): 125–53.

13. "The wireless produces, without the need for experiments or association tests, words in statistical dispersion, freed from reference and addresses (except for the numerical identification of stations)." Friedrich Kittler, "Benn's Poetry—'A Hit in the Charts': Song under Conditions of Media Technologies," *Substance: A Review of Literary Theory and Criticism,* no. 61 (1990): 13.

14. This is perhaps what Nikola Tesla had in mind as he envisioned a code designed for clarity, a hyperhermeneutical code with which to greet our cosmic neighbors: "Brethren! We have a message from another world, unknown and remote. It reads: one . . . two . . . three." Kahn, *The Codebreakers,* 952. Or consider the German military cipher, "Für Gott," so called "because all messages bore that prefix to show that they were for the German wireless station whose call letters were GOD" (310).

15. Manuel De Landa, *War in the Age of Intelligent Machines* (New York: Zone Books, 1991), 180. "When radio replaced the telegraph, it forced the development of a different approach, since messages were not carried by wires anymore, but released directly into the electromagnetic spectrum. Instead of wiretapping techniques, the new media forced the development of hypersensitive antennae to snatch very faint signals 'out of the ether'" (180).

16. "I would say that when a voice has a localizable place in a social field, it is no

longer heard; when one does hear it, it is the voice of a ghost, a voice that is seeking its place like a will-o'-the-wisp; when you hear a voice over the telephone, on the radio, you don't see where it comes from; as for the one who seems to be the bearer of the voice, you don't know with a certain knowledge either where or who he or she is; the bearer of him or herself does not know either." Jacques Derrida, "Dialanguages," in *Points . . .: Interviews* (Stanford, CA: Stanford University Press, 1995), 135.

17. His greatest successes were Radio Mondial and the Communist station Radio Humanitè. "The task of the secret transmitter, from now on, is to use every means to create a mood of panic in France. To that end it must work on a wholly French platform and display the greatest indignation and alarm in protesting against the omissions of the French government. In particular, the rumors buzzing around France are to be picked up and developed." Willi A. Boelcke, ed. and trans., *The Secret Conferences of Dr. Goebbels: The Nazi Propaganda War, 1939–43* (New York: E. P. Dutton, 1970), 42. Utilizing a Communist transmitter also worked well: "The communist transmitter, likewise, must no longer operate with Soviet slogans, but must put forward arguments of purely social agitation. . . . Two of the secret transmitters are to keep their 'location' in occupied territory and may put out reports such as the entry of the German troops into Paris" (53–54).

18. Riegel, *Mobilizing for Chaos,* 29.

19. Unfortunately, no biography of Alexanderson exists, though he is one of the most important figures of post-Marconi wireless technology. For a cursory look at his life, see James E. Brittain's examination of alternators for a history of electricity, "The International Diffusion of Electrical Power Technology, 1870–1920," *Journal of Economic History* 34, no. 1 (March 1974): 108–21.

20. Kern, *Culture of Time and Space,* 308. Kern is surely right in locating the fabrication of speed before the war as one of the chief ways in which distance was experienced during it, but this optic drives him too often to conflating a transmission medium such as the wireless with the same distance-shrinking machines.

21. See where British soldiers located their commanding officer (CO) in the wartime song "The Old Battalion": "If you want to find the CO, / I know where he is, I know where he is; / If you want to find the CO, / I know where he is: / He's down in the deep dugouts." The old battalion is "on the old barbed wire, / I've seen them, I've seen them." Paul Fussell, ed., *The Norton Anthology of War* (New York: W. W. Norton & Company, 1991), 84–85.

22. "[T]raditional warfare could simply not do without the physical presence of chiefs or seconds-in-command." Friedrich Kittler, "Media Wars: Trenches, Lightning, Stars," in *Literature Media Information Systems,* ed. John Johnston (Amsterdam: Overseas Publishers Association, 1997), 117.

23. Albert Speer, *Inside the Third Reich,* trans. Richard and Clara Winston (New York: Avon Books, 1970), 653–54.

24. Alessandro Silvestri, *La tecnica del secolo* (Milan: Casa Editrice Dr. Francesco Vallardi, 1956), 131.

25. Jacques Lacan, "The Creative Function of Speech," in *The Seminar of Jacques Lacan,* vol. 1, *Freud's Papers on Technique, 1953–1954,* 240; emphasis in original.

26. Ibid.

27. See Pound's speech "America Was Promises": "I do what I can to keep an even tone of voice; now when I drop my voice, they turn on more current." In *"Ezra Pound Speaking": Radio Speeches of World War II,* ed. Leonard W. Doob, Contributions in American Studies, no. 37 (Westport, CT: Greenwood Press, 1978), 387.

28. What I have in mind in the distinction between meaning and recognition is something like the categories of attribution and information for systems theory. "If, in deconstruction, everything is motivated by the linguistically induced differentiality of presence/absence, the equivalent of this in systems theory is the difference between information and utterance inherent in communication. Here, information refers to the world, and the utterance to the speaker, and this difference is permanently displaced. And this is precisely the effect of attribution." Dietrich Schwanitz, "Systems Theory and the Difference between Communication and Consciousness: An Introduction to a Problem and Its Context," *MLN* 111 (April 1996): 489.

29. Everett P. Gordon, *Acquiring the Code: A Text on Key Manipulation and the Proper Method for Acquiring the International Radio Code* (Everett P. Gordon, 1920), 3. Of particular difficulty were the letters "C" and "N" for the novice wireless operator, leading to Exercise 5: "While practicing the following exercises bear in mind that there must be no division or break made in any one letter. Weave the constituents of the letter so closely together that there is no doubt as to what is meant. Practice the letter 'C' very carefully. It is very easy to execute this letter as a cross between 'C' and the letter 'N' sent twice. Guard against this" (6).

30. Ibid., 15.

31. Tom Lewis, *Empire of the Air: The Men Who Made Radio* (New York: Edward Burlingame Books, 1991), 97.

32. "Are we going to say that semantics is peopled, furnished with the desire of men? What is certain is that it is us who introduce meaning. In any case it is certain for a great number of things." Jacques Lacan, "Psycho-Analysis and Cybernetics, or on the Nature of Language," in *The Seminar of Jacques Lacan,* vol. 2, *The Ego in Freud's Theory and in the Technique of Psychoanalysis, 1954–55,* 305.

33. Rudolf Arnheim, *Radio,* trans. Margaret Ludwig and Herbert Read (London: Faber & Faber, 1936), 143.

34. Friedrich A. Kittler, *Gramophone, Film, Typewriter,* trans. Geoffrey Winthrop-Young and Michael Wutz (Stanford, CA: Stanford University Press, 1999), 3.

35. Gordon, *Acquiring the Code,* 10. He goes on to identify the two processes whereby a transmission is received: "In the sound method we conceive each signal as an entirety, the cadence peculiar to each character distinguishing it from all others. Thus when a sound is heard there is no occasion for a visual interpretation of it."

36. Arnheim, *Radio,* 136.

37. Warren Weaver, "Some Recent Contributions to the Mathematical Theory of Communication," in *The Mathematical Theory of Communication,* by Claude E. Shannon (Chicago: University of Illinois Press, 1963), 18.

38. See blind broadcaster Arthur T. Cushen's chapter "My Darkest Hour" for how blindness makes possible greater speed in transmission: "The material is then prepared

into the Braille script, but, owing to the fact that Braille takes five times more paper than print, the actual details are kept to a minimum with country, frequency, schedule and a few other details noted, while the rest of the material is adlib." Arthur T. Cushen, *The World in My Ears* (n.p., 1979), 52.

39. "If the incoming wave of, say, 51,000 cycles is mixed with a wave created by an oscillator within the receiver of 50,000 cycles, the result is a third, audible wave of 1,000 cycles." Lewis, *Empire of the Air*, 133.

40. *Radio Pamphlet No. 40*, December 10, 1918, Signal Corps, U.S. Army (Washington, DC: Government Printing Office, 1919), 228.

41. Lewis, *Empire of the Air*, 134.

42. "Radio is superior to nature; it contains more and is capable of being varied." Gottfried Benn, *Gesammelte Werke*, ed. Dieter Wellershoff, vol. 2 (Wiesbaden: Limes Verlag, 1959–61), 182; quoted in Kittler, "Benn's Poetry," 12.

43. E. A. Armstrong, "The Super-Heterodyne Receiver: An Account of Its Origin, Development, and Some Recent Improvements," *Wireless World*, November 24, 1924, at http://home.luna.nl/~arjan-muil/radio/superheterodyne.htm.

44. As more and more stations began broadcasting, interference grew acute. Often the result was the "heterodyne tone," which arose when the signals of two stations were too close in frequency. "For example, if one station were on 833 khz, and the other on 830 khz, the resulting heterodyne tone would be 3 khz, which [following Fessenden's theorem] is the difference between the two station frequencies." Thomas H. White, "Building the Broadcast Band," January 1, 1998, at http://www.oldradio.com/archives/general/buildbcb.html.

45. "Reformers, monetary reformers who haven't even yet arrived at the concept of linotype government, cannot expect to rule in a radio age." Ezra Pound, *Guide to Kulchur* (New York: New Directions, 1970), 241.

46. Forrest Read, *Pound/Joyce: The Letters of Ezra Pound to James Joyce, with Pound's Essays on Joyce* (London: Faber and Faber, 1967), 13.

47. Ibid.

48. "Nietzsche's Dane from Copenhagen was Malling Hansen, pastor and teacher of the deaf and dumb, whose 'writing ball' of 1865 or 1867 was designed for use only by the blind." Friedrich Kittler, *Discourse Networks, 1800/1900,* trans. Michael Metteer (Stanford, CA: Stanford University Press, 1990), 193. Cf. Derrida's "I am writing to you now on the typewriter, it can be felt." Jacques Derrida, *The Post Card: From Socrates to Freud and Beyond,* trans. Alan Bass (Chicago: University of Chicago Press, 1987), 249.

49. T. S. Eliot, *The Waste Land: A Facsimile and Transcript of the Original Drafts Including the Annotations of Ezra Pound,* ed. Valierie Eliot (London: Faber and Faber, 1971), 55. For a history of Pound and Eliot's literary relations, see Erik Svamy's *"The Men of 1914": T. S. Eliot and Early Modernism* (New York: Open University Press, 1988); and J. J. Wilhelm's *Ezra Pound in London and Paris, 1908–1925* (University Park: Pennsylvania State University Press, 1990).

50. De Rachewiltz, *Discretions*, 139. Tennis as a metaphor for Pound's various dispatches appears once more when he decides to begin broadcasting: "The idea of broadcasting had originally been suggested to him by C. H. Douglas in 1935. After Pound

returned from his 1939 visit to the United States, a Nazi officer with whom he played tennis encouraged him to pursue it. Pound then approached the Ministry of Popular Culture and made his offer." G. Fuller Torrey, *The Roots of Treason: Ezra Pound and the Secret of St. Elizabeth's* (New York: McGraw-Hill, 1984), 157. (I have found, however, no confirmation of Torrey's anecdote.) Often, tennis stands in for the symbolic sport of data retrieval as in the following passages: "As you would not seriously consider a man's knowledge of tennis until he either could make or had some sort of show in a tournament, so we can assume that until a man can actually control a given set of procedures, there must be many elements in them of which he has but an imperfect knowledge." Ezra Pound, "Date Line," in *Literary Essays* (London: Faber and Faber, 1954), 74.

51. "Apparently, Amprim or some other FBI agent examined Pound's typewriter and found that the letter 't' was out of alignment; in this way the FBI traced some of his war-time writing." Noel Stock, *The Life of Ezra Pound* (New York: Avon Books, 1970), 525. Regarding Pound's handwriting, see the report of Dr. Keeney, one of Pound's psychiatrists, who believed that Pound's diminished ability to write legibly was a cause for concern: "He still sends and receives quite a lot of mail and his cryptographic handwriting is becoming more and more illegible." Generally, Pound wrote at St. Elizabeth's with a pencil, having refused the offer of a typewriter or a pen. Dr. Keeney, minutes of weekly session with Ezra Pound, March 12, 1948, Ezra Pound–St. Elizabeth's file, U.S. National Archives, Washington, DC.

52. Quoted in Kittler, *Gramophone, Film, Typewriter,* 191.

53. Robert Lincoln O'Brien, "Machinery and English Style," *Atlantic Monthly* 94 (1904): 470–71.

54. De Rachewiltz, *Discretions,* 256. Compare Pound's instructions in *Guide to Kulchur:* "The bare wrong phrase carries a far heavier charge of meaning than any timorous qualification as 'We admit that beauty can't be hustled, it cannot be scamped for time'" (129).

55. Niklas Luhmann, *Social Systems,* trans. by John Bednarz, Jr. (Stanford, CA: Stanford University Press, 1995), 139.

56. De Rachewiltz, *Discretions,* 257.

57. Cf. Pound's "caesarian operation" in *The Waste Land* (letter to Eliot, January 24, 1922) and Eliot's scribbled fragments: "whereupon Eliot—reluctant to let the transformation theme go—scribbled a new fragment on his 'British Bond' paper . . . Pound looked critically at 'Death by Water' and ordered Eliot to type it. He drew a thick line through the central 'London' fragment in the 'The Fire Sermon,' and cancelled references to churches, Michael Paternoster and St. Mary Woolnoth." Lyndall Gordon, "The Waste Land Manuscript," *American Literature* 45 (January 1974): 568.

58. Rolf Stümpel, *Vom Sekretär zur Sekretärin: Eine Ausstellung zur Geschichte der Schreibenmaschine und ihrer Bedeutung für den Beruf der Frau im Büro* (Mainz, Germany: Gutenberg Museum, 1985), 12; quoted in Kittler, *Gramophone, Film, Typewriter,* 290.

59. See note 5 above.

60. All references to the *Cantos* are to Faber and Faber's 1975 Revised Collected edition. For a history of the various editions, see Peter Stoicheff's chapter on the final edition of the *Cantos* in *A Poem Containing History: Textual Studies in "The Cantos,"* ed.

Lawrence Rainey (Ann Arbor: University of Michigan Press, 1997), and Richard Taylor's "Reconstructing Ezra Pound's Cantos: Variorum Edition—Manuscript Archive—Reading Text," in *Ezra Pound and America,* ed. Kaye, 132–48.

61. For Gertrude Stein, Pound was "a village explainer, excellent if you were a village, but if you were not, not." Gertrude Stein, *Autobiography of Alice B. Toklas* (New York: Literary Guild, 1934), 200; quoted in Whittier-Ferguson, *Framing Pieces,* 4.

62. Ezra Pound, "The Serious Artist," in *Literary Essays,* 49.

63. In his *Shapes of Power,* Bruce Fogelman finds Pound attempting to maintain the poem's coherence (and failing) in the use of juxtaposition of verbal units: parataxis equals compression. John H. Zammito cuts to the chase by linking coherence to voice: "What we admire is the scope of the coherence, the polyphony of voices meaningfully situated within the work. . . . Coherence is nowhere in the literal text. It is, however, the name humans give to that sense of meaningful closure, of fittingness." "Are We Being Theoretical Yet? The New Historicism, the New Philosophy of History, and 'Practicing Historians,'" *Journal of Modern History* 65 (December 1993): 813.

64. Cf. Pound's reasons for the speed of Joyce's prose: "This variegation of dialects allows Joyce to present his matter, his tones of mind, very rapidly." Ezra Pound, "Ulysses," in *Literary Essays,* 404.

65. "A voice detaches itself, that is its way of 'attaching itself.' In any case, if there is any 'locus' where the figure of connection (attachment/detachment, binding/unbinding) no longer offers the least security, then surely it is an atopical voice, this madness of voice." Jacques Derrida, "Voice II," in *Points . . .: Interviews,* 160–61.

66. "Wherever one meets the 'content' concept, it is reasonably certain that there has been insufficient structural analysis. Phonetic writing and printing, for example, have content only in the sense that they 'contain' another medium, namely speech. But since the origin of writing, the simultaneous presence of the medium of speech, albeit in low definition, has fostered this habit of dichotomy and content-postulating, which in fact obscures major components in the situation in which we must deal." Marshall McLuhan, "Effects of the Improvements of Communication Media," *Journal of Economic History* 20 (December 1960): 572. Cf. Derrida and media wars: "There would only be *'facteurs'* and therefore no *vérité.* Only 'media' take this into account in every war against the media. The immediate will never be substituted for them, only other frameworks and other forces." Derrida, *The Post Card,* 194.

67. Kahn, *The Codebreakers,* 559.

68. "If we admit that the writing process is not only the transposition of a text which existed originally in the writer's mind, then the role of the inscriptive gesture in the writing process has been generally and markedly underestimated, particularly by linguistics and semiology." Serge Tisseron, "All Writing Is Drawing: The Spatial Development of the Manuscript," *Yale French Studies* 84 (1994): 29.

69. "But beyond theoretical mathematics, the development of the *practical methods* of information retrieval extends the possibilities of the 'message' vastly, to the point where it is no longer the 'written' translation of a language, the transporting of signified which could remain spoken in its integrity. It goes hand in hand with an extension of phonography and of all the means of conserving the spoken language, of making it func-

tion without the presence of the speaking subject." Jacques Derrida, *Of Grammatology,* Corrected Edition, trans. Gayatri Chakravorty Spivak (Baltimore, MD: Johns Hopkins University Press, 1997), 10; emphasis in original.

70. Kittler, *Gramophone, Film, Typewriter,* 8.

71. Pound, "The Serious Artist," in *Literary Essays,* 42.

72. Pound, "The Teacher's Mission," in *Literary Essays,* 58.

73. Pound, "Dr. Williams' Position," in *Literary Essays,* 396; emphasis in original.

74. Pound, "The Serious Artist," 44.

75. "At one time or another, most of the Modernist writers seem to have had recourse to the vocabularies of science and technology to define their practices and objectives. That habit of reference seems a commonplace of the period, but it is one which still raises awkward questions . . .; and how could the pure functionalism of technology provide an accurate index of what was being sought in poetic language?" Peter Nicholls, "Machines and Collages," *Journal of American Studies* 22 (1988): 275. Nicholls assumes something like a poetic language at the disposal of the Modernist poet but elides the mediated "pure functionalism" of poetic language before the advent of the machine. Still, his notion of a "linguistic and imaginative surplus" situated between representation and the materiality of the medium recalls the effects of wireless installation in Marinetti.

76. Jessica Burstein, "Waspish Segments: Lewis, Prosthesis, Fascism," *Modernism/ Modernity* 4 (April 1997): 153. "Tied historically to the development of modern warfare, cinema gave us new perspectives, literally, on 'phenomena of the crowd' such as war and the street demonstration. It shapes a new body with which to experience violence, positioning us, anaesthetizing us, sensitizing us."

77. In *Guide to Kulchur,* Pound describes the nexus more explicitly: "Nothing exists. All is from a rain of atoms. If anything did exist it wd. be ununderstandable and if understandable wd. be incommunicable. . . . All knowledge is built up from a rain of factual atoms" (98).

78. Ibid., 50.

79. Kittler, "Benn's Poetry," 18.

80. Cornelio Fazio, Diagnosis Sheet, Ezra Pound–St. Elizabeth's file, U.S. National Archives, Washington, DC. It is unclear how Dr. Fazio's examination records of such a late date arrived in the file. The most likely explanation appears to be some sort of correspondence between Fazio and Pound's doctors at the hospital on the occasion of Pound's nervous collapse. For an account of Pound's stay at St. Elizabeth's, see Piero Sanavío, *La gabbia di Ezra Pound* (Milan: Scheiwiller, 1986), and the indispensable David M. Gordon, ed., *Ezra Pound and James Laughlin: Selected Letters* (New York: W. W. Norton, 1994).

81. Dr. Kavka, family history of Ezra Pound, January 24, 1946, Ezra Pound–St. Elizabeth's file.

82. Dr. Stevens, case history of Ezra Pound, March 31, 1946, Ezra Pound–St. Elizabeth's file.

83. Ezra Pound, letter to James Laughlin; quoted in Stock, *Life of Ezra Pound,* 483.

84. Pound was first admitted to St. Elizabeth's Hospital on December 21, 1945, as a prisoner of the U.S. government. He was judged of unsound mind on February 13, 1946.

85. "Do you have a long-playing phonograph machine down there?" James Laughlin asks Dr. Overholser, Pound's principal doctor. "If you do, I'll send you a copy of the record of Ezra reading his poetry which has just come out" (letter, April 29, 1960, Ezra Pound–St. Elizabeth's file). "I do have a long-playing Magnavox," Dr. Overholser responds, "and should be delighted to have a copy of the record of Ezra reading his poetry" (letter, June 27, 1960, Ezra Pound–St. Elizabeth's file).

86. Pound, *Guide to Kulchur*, 60.

87. Weaver, "Some Recent Contributions to the Mathematical Theory of Communication," 19. "It is thus clear where the joker is in saying that the received signal has more information. Some of this information is spurious and undesirable and has been introduced via the noise. To get the useful information in the received signal we must subtract out this spurious portion."

88. "A medium is a medium is a medium." Kittler, *Discourse Networks*, 265.

89. Whittier-Ferguson, *Framing Pieces*, 16. A further difficulty lies in the paradoxical relationship fascism holds with the "openness" of the poem. "The unifying impulse in the Cantos can be partly accounted for by Pound's attraction to fascism, which in turn signifies a covert revulsion for his modernist poetic practice. The directness and unity valued by Pound's political ideology are violated by the poem's openness and indeterminacy. These characteristics are related to Pound's interest in history." The statement is problematic for a number of reasons, not least its failure to account for fascist appropriation of Futurist practices. Cordell De. K. Yee, review of *A Calculus of Ezra Pound: Vocations of the American Sign*, by Philip Luberski, in *American Literature* 65 (December 1993): 804.

90. David Bathrick, "Making a National Family with the Radio: The Nazi *Wunschkonzert*," *Modernism/Modernity* 4 (March 1997): 118.

91. Speer, *Inside the Third Reich*, 653–54.

92. Delmore Schwartz, "Ezra Pound's Very Useful Labors," *Poetry* (March 1938), quoted in *Ezra Pound: The Critical Heritage*, ed. Eric Homberger (London: Routledge & Kegan, 1972), 315. Schwarz rephrases his argument on the following page: "But most of all, literary practice benefits by the effort of the Cantos to digest a great many diverse elements, and to speak, in one poem, of many different *kinds* of things" (316; emphasis in original).

93. W. D. Snodgrass, *Hudson Review* 12 (Spring 1960); quoted in *Ezra Pound: The Critical Heritage*, ed. Homberger, 463.

94. Pound's fascination with Mussolini and fascism, however, began earlier: "I personally think extremely well of Mussolini. If one compares him to American presidents (the last three) or British premiers, etc., in fact one can NOT without insulting him." Letter to Harriet Monroe, November 30, 1926, in *The Letters of Ezra Pound, 1907–1941*, ed. D. D. Paige (London: Faber and Faber, 1951), 279.

95. Paul Fussell, *The Great War and Modern Memory* (London: Oxford University Press, 1975), 106.

96. G. Singh, ed., *The Sayings of Ezra Pound* (London: Duckworth, 1994), 63.

97. George P. Elliott, "Poet of Many Voices," in *Ezra Pound: A Collection of Critical Essays*, ed. Walter Sutton (Englewood Cliffs, NJ: Prentice-Hall, 1963), 152.

5. Pound's Marconigrams

1. Quoted in Mary de Rachewiltz, "Fragments of an Atmosphere," in *Agenda,* no. 17 (1979): 161.

2. For the exceptions, see Shef Rogers, "How Many Ts Had Ezra Pound's Typewriter," *Studies in Bibliography: Papers of the Bibliographical Society of the University of Virginia* 49 (1996): 277–83; Max Nänny, "The Oral Roots of Ezra Pound's Methods of Quotation and Abbreviation," *Paideuma: A Journal Devoted to Ezra Pound Scholarship* 8 (1979): 381–87; and Leonard W. Doob's introduction to *"Ezra Pound Speaking": Radio Speeches of World War II* (Westport, CT: Greenwood Press, 1978).

3. "Had he been among the delighted New Yorkers, Guyau would have found empirical proof that frequency modulation is indeed the technological correlative of attention." Friedrich A. Kittler, *Gramophone, Film, Typewriter,* trans. Geoffrey Winthrop-Young and Michael Wutz (Stanford, CA: Stanford University Press, 1999), 35.

4. Ezra Pound, *Guide to Kulchur* (New York: New Directions, 1970).

5. Rudolf Arnheim, *Radio,* trans. Margaret Ludwig and Herbert Read (London: Faber & Faber, 1936), 156.

6. Omar Pound and Robert Spoo, eds., *Ezra and Dorothy Pound: Letters in Captivity, 1945–46* (New York: Oxford University Press, 1999); and Mary de Rachewiltz, *Discretions* (Boston: Little, Brown, 1971).

7. Kittler, *Gramophone, Film, Typewriter,* 22–23.

8. Under this rubric, see Adam Parkes, "Ezra Pound as Censor," *Centennial Review* 43 (1999): 259–88; Ulisse Belotti, "Ezra Pound e il proiettile magico," *Confronto Letterario: Quaderni del Dipartimento di Lingue e Letterature Straniere Moderne dell'Università di Pavia* 14 (1997): 311–28; Jonathan Gill, "The Devil Box: Radio Space and Radio Subject in Ezra Pound's Broadcasts," *TheoryBuffalo.edu* 1 (Summer 1995): 135–49; Conrad L. Rushing, "'Mere Words': The Trial of Ezra Pound," *Critical Inquiry* 14 (1987): 111–33; and Philip Furia, *Pound's Cantos Declassified* (University Park: Pennsylvania State University Press, 1985).

9. Jeffrey Schnapp, "Politics and Poetics in Marinetti's *Zang Tumb Tuuum,*" *Stanford Italian Review* 5 (Spring 1985): 79.

10. Arnheim, *Radio,* 82.

11. Ibid., 100.

12. Ibid., 100–101.

13. Ibid., 278.

14. Ibid., 110.

15. Ibid., 99.

16. Ibid., 143.

17. Ibid., 173. "We demand that nothing should be brought in just for its own sake, but everything must be essential to the effect produced. The frame of the picture must not only be an accidental limit (because even the largest picture must come to an end sometime); in a play every character must have his definite rôle, in a piece of music no instrument should merely have the task of filling in the parts. It is just this indispensability of all its parts that differentiates the work of art from reality" (133–34).

18. Ibid., 121.

19. Ibid., 195.

20. Ibid., 152–53.

21. Ibid., 154.

22. Ibid., 155, 156, 242.

23. Arnheim was forced out of Germany soon after the publication of his first work, *Film as Art,* due to his Jewish ancestry. From there he moved to Rome where he wrote *Radio.* On Arnheim's life and influence, see "The Work of Rudolf Arnheim," *Salmagundi* 78–79 (Spring–Summer 1988): 43–143; "Essays in Honor of Rudolf Arnheim," *Journal of Aesthetic Education* 27 (Winter 1993); "Rudolf Arnheim," *Dictionary of Art,* Jane Turner (New York: Grove Dictionaries, 1996); and Rudolf Arnheim, *My Life in the Art World* (Ann Arbor: School of Art, University of Michigan, 1984). For these citations, I am indebted to http://www.mitpress.com/e-journals/Leonardo/isast/articles/arnheim.html, which also contains a brief but useful biography.

24. Arnheim, *Radio,* 244.

25. Ibid., 27.

26. Ibid., 28.

27. As Cocteau does in his insight into the hermeneutic appetite. "This craving to understand (when the world that people inhabit and acts of God are apparently incoherent, contradictory and incomprehensible), this craving to understand, I say, shuts them off from all the great and exquisite precisions that art deploys in the solitudes where men no longer try to understand, but to feel." Jean Cocteau, *The Art of Cinema,* trans. Robin Buss (New York: Marion Boyars, 1988), 42.

28. For Arnheim as for Derrida, the ear is the site through which the word travels to meaning. It must be dressed properly. "The words of a radio play should not go about in acoustic hair-shirts, they should shimmer in all their tone-colours, for the way to meaning of the words lies through the ear" (Arnheim, *Radio,* 29). "This, my love, me: the last photomaton. I will have written to you, written also in every code, loved according to every genre. All colors, all tones are ours." Jacques Derrida, *The Post Card: From Socrates to Freud and Beyond,* trans. Alan Bass (Chicago: University of Chicago Press, 1987), 109.

29. "With expensive and disloyal mercenaries out of the way he could afford to gamble armies in decisive clashes without worrying about shortages of reserves, and without having to fear troop desertion while pursuing a defeated enemy." Manuel De Landa, *War in the Age of Intelligent Machines* (New York: Zone Books, 1991), 182.

30. A full accounting of wireless switches is beyond my present means, though the reader is directed to Marinetti's radio manifesto, "La radia," for a list of some of the most important. The third paragraph in particular merits attention: "Overcome patriotism 'with a more fervent patriotism thus transformed into authentic religion of the Fatherland warning to Semites that they should identify with their various fatherlands if they do not wish to disappear.'" F. T. Marinetti, "La radia," in *Wireless Imagination: Sound, Radio, and the Avant-Garde,* ed. Douglas Kahn and Gregory Whitehead, trans. Stephen Sartarelli (Cambridge, MA: MIT Press, 1992).

31. Arnheim, *Radio,* 49.

32. Ibid., 82.

33. Arnheim describes the antagonism required to reach hearts and heads in "Dynamics," a chapter from his opus *Art and Visual Perception*. "Perception reflects an invasion of the organism by external forces, which upset the balance of the nervous system. A hole is torn in a resistant tissue. A struggle must result as the invading forces try to maintain themselves against the physiological field forces, which endeavor to eliminate the intruder or at least reduce it to the simplest possible pattern." Rudolf Arnheim, *Art and Visual Perception: A Psychology of the Creative Eye* (Berkeley: University of California Press, 1954), 438.

34. Arnheim, *Radio,* 119.

35. Kittler, *Gramophone, Film, Typewriter,* 98.

36. Arnheim, *Radio,* 79.

37. The wireless and gramophone limit the divergent frequencies of spoken language. See Goebbels's repeated attempts to fix German pronunciation: "The radio, the Minister says, must speak the purest, clearest and most dialect-free German because it speaks to the whole nation. As a matter of principle, there must be a certain standard language just as there is a standard orthography. What Luther's translation of the Bible has done for written German, we on the radio must do for German speech: fix a standard language which, even if it does not sweep away the dialects, is valid throughout the Reich." Willi A. Boelcke, ed., *The Secret Conferences of Dr. Goebbels: The Nazi Propaganda War, 1939–43,* trans. Ewald Osers (New York: D. P. Dutton, 1970), 140.

38. Arnheim, *Radio,* 128.

39. David Kahn, *The Codebreakers: The Story of Secret Writing* (London: Weidenfeld and Nicholson, 1967), 554.

40. "Broadcasting of weightless material came about for the purpose of the mass transmission of records: in 1921 in the United States, in 1922 in Great Britain, and in 1923 in the German Reich" (Kittler, *Gramophone, Film, Typewriter,* 94). There are numerous histories of radio broadcasting available. In the American context, the best include James Wood, *History of International Broadcasting,* History of Technology Series, no. 19 (London: P. Peregrins, 1992); Christopher H. Sterling and John M. Kittross, *Stay Tuned: A Concise History of American Broadcasting* (Belmont, CA: Wadsworth, 1978); and Erik Barnouw, *A Tower in Babel: A History of Broadcasting in the United States,* vol. 1 (New York: Oxford University Press, 1966). For Great Britain, see Hanno Hardt, *In the Company of Media: Cultural Constructions of Communication, 1920s–1930s,* Critical Studies in Communication and in the Cultural Industries (Boulder, CO: Westview Press, 2000); and Jennifer R. Doctor, *The BBC and Ultra-Modern Music, 1922–1936: Shaping a Nation's Tastes,* Music in the Twentieth Century (New York: Cambridge University Press, 1999). Gianni Isola's *Abbassa la radio, per favore . . . : Storia dell'ascolto radiofonico nell'Italia fascista* (Florence: La Nuova Italia, 1990) is a fine introduction to the Italian context, as is Giorgio Maioli's *I giorni della radio: A cent'anni dall'invenzione di Guglielmo Marconi* (Bologna: Re Enzo, 1994). For a reading of amplification and fascism, see Jeffrey T. Schnapp's *Staging Fascism: 18 BL and the Theater of Masses for Masses* (Stanford, CA: Stanford University Press, 1996).

41. Arnheim, *Radio,* 126.

42. Ibid., 131.

43. Kittler, *Gramophone, Film, Typewriter*, 94–105.

44. Arnheim, *Radio*, 56, 189.

45. Ibid., 141.

46. Ibid., 140.

47. Ibid., 122.

48. David E. Wellbery, foreword to *Discourse Networks, 1800/1900*, Friedrich A. Kittler, trans. Michael Metteer (Stanford, CA: Stanford University Press, 1990), xxv.

49. Pound, *Guide to Kulchur*, 51.

50. Ibid., 57.

51. Ibid., 28.

52. Ibid.

53. The relationship the production of copy entertains with individual difference ought to be recalled. "The authenticity of a signature is thereby allied to prior modes of authentication; there is an implicit recognition that the pedagogy of copy produces the self-imitating individual and that it does not guarantee individual difference." Jonathan Goldberg, *Writing Matter: From the Hands of the English Renaissance* (Stanford, CA: Stanford University Press, 1990), 246.

54. Pound, *Guide to Kulchur*, 29.

55. Apparently Pound had other communication systems at his disposal. Describing their sitting with a portraitist, de Rachewiltz observes her father. "His eyes and smile conveyed that he would rather she did one of me first. He was capable of saying a great deal by just blinking and smiling and shifting his weight from one foot to the other" (de Rachewiltz, *Discretions*, 111).

56. Pound, *Guide to Kulchur*, 53.

57. "On the one hand, there was speech, and the handwriting animated by it: on the other there was automated and automatic writing. Where speech-writing is intimate and expressive, machine-writing is abstract and impersonal." Steven Connor, paper delivered at the "Modernism and the Technology of Writing" conference, March 26, 1999, at http://www.bbk.ac.uk/eh/eng/skc/mod-hand.htm.

58. "It is the same with language, which only leaves us the choice of either retaining words while losing their meaning or, vice versa, retaining meaning while losing the words. Once storage media can accommodate optical and acoustic data, human memory capacity is bound to dwindle." Kittler, *Gramophone, Film, Typewriter*, 10.

59. Pound, *Guide to Kulchur*, 57.

60. Cf. Pound's early poem "The Cry of the Eyes": "Rest Master, for we be aweary, weary / And would feel the fingers of the wind / Upon these lids that lie over us / Sodden and lead-heavy . . . Free us, for we perish / In this ever-flowing monotony / Of ugly print marks, black / Upon white parchment." In Ezra Pound, *Personae: The Collected Poems of Ezra Pound* (New York: New Directions, 1965), 24.

61. Pound, *Guide to Kulchur*, 59.

62. Ibid., 44–45.

63. Jacques Lacan, *Television*, trans. Denis Hollier, Rosalind Krauss, and Annette Michelson (New York: W. W. Norton, 1990), 37.

64. Pound, *Guide to Kulchur*, 44.

65. Ezra Pound, *Machine Art and Other Writings,* ed. Maria Luisa Ardizzone (Durham, NC: Duke University Press, 1996).

66. The full citation reads: "There is an essential link which must be made right away—when you draw a rabbit out of a hat, it's because you put it there in the first place. Physicists have a name for this formulation, they call it the first law of thermodynamics, the law of the conservation of energy—if there's something at the end, just as much had to be there at the beginning." Jacques Lacan, "The Circuit," in *The Seminar of Jacques Lacan,* vol. 2, *The Ego in Freud's Theory and in the Technique of Psychoanalysis, 1954–55,* trans. Sylvana Tomaselli (New York: W. W. Norton, 1988), 81.

67. Martin Gardner, *Logic Machines, and Diagrams* (Chicago: University of Chicago Press, 1982), 127; quoted in de Landa, *War in the Age of Intelligent Machines,* 145.

68. De Landa, *War in the Age of Intelligent Machines,* 146.

69. "The electrical signals which pass along the nervous systems of animals and men, both from the sense organs (receptors) and to the controlled organs and muscles (effectors), take the form of triggered pulses which are either *on* or *off;* there is no half measure." Colin Cherry, ed., *Information Theory: Symposium on Information Theory,* (New York: Academic Press, 1956), 35, quoted in *Handbook of Semiotics,* Winfred Nöth (Bloomington: Indiana University Press, 1990). See also V. Pekelis's summary of a Norbert Wiener lecture of 1960: "He said, among other things, that electric oscillations with a frequency of around 10 hertz have a maximum energy compared with all the other oscillations occurring in the human brain. The pattern of maximum-energy signals consists of sharp peaks alternating with valleys. The peak frequency is termed alpharhythm." V. Pekelis, "Several Episodes from My Personal Experience with Telepathy," in *Radiotext(e),* ed. Neil Strauss (New York: Semiotext(e), 1993), 327–28.

70. Marshall McLuhan makes the same argument in an unpublished letter to Pound of June 2, 1951: "Also I'm interested in such analogies with modern poetry as that provided by the vacuum tube. The latter can tap a huge reservoir of electrical energy, picking it up as a very weak impulse. Then it can shape it and amplify it to major intensity. Technique of allusion as you use it (situational analogies) seems comparable to this type of circuit. Allusion not as ornament but as precise means of making available total energy of any previous situation or culture. Shaping and amplifying it for current use." Edwin Barton, "On the Ezra Pound Marshall McLuhan Correspondence," at http://www.epas .utoronto.ca/mcluhan-studies/v1_iss1/1.

71. Jacques Lacan, "Sign, Symbol, Imaginary," in *On Signs,* ed. Marshall Blonsky (Baltimore, MD: Johns Hopkins University Press, 1985), 207.

72. Ibid.

73. "Because truth flows naturally from general principles (axioms) to particular statements (theorems), it is a relatively simple task to create a set of rules (or mechanical device) to perform this operation." De Landa, *War in the Age of Intelligent Machines,* 144. See the calculations Derrida undertakes in *The Post Card:* "Trrrrr, *je trame,* I weave, *je trie,* I sort, I treat, I traffic, I transfer, I intricate, I control, *je filtre,* I filter—and as I have done so often on leaving, I am leaving the note in the box" (232).

74. Pound, *Guide to Kulchur,* 44.

75. Ibid., 51.

76. "'Trasmettere la veritá, da parte di Dio, all'uomo' (1321, Dante)." *Dizionario etimologico della lingua italiana* (1988), s.v. "Rivelare." See as well Pound's etymology for "symptom": "Not to minimize the danger of cocacola'd minds, BUT to kick the pussilanimous *[sic]* who do NOT observe the symptoms [ideogram often translated as 'omens']." Ezra Pound, St. Elizabeth's, to Olivia Rossetti Agresti, January 7, 1952, in *"I Cease Not to Yowl": Ezra Pound's Letters to Olivia Rossetti Agresti,* ed. Demestres P. Tryphonopoulous and Leon Surette (Urbana: University of Illinois Press, 1998), 83.

77. Lacan, "Sign, Symbol, Imaginary," 209.

78. Pound, *Guide to Kulchur,* 123.

79. Kittler, "The World of the Symbolic," in *Literature Media Information Systems: Friedrich A. Kittler Essays* (Amsterdam: Overseas Publishers Association, 1997), 145.

80. In this regard, compare Derrida's musings on nonlinear writing from twenty-five years ago: "The end of linear writing is indeed the end of the book. . . . It is less a case of confiding new writings to the envelope of the book than of finally reading what wrote itself between the lines in the volumes. That is why, beginning to write without the line, one begins to reread past writing according to a different organization of space. . . . What is thought today cannot be written according to the line and the book, except by imitating the operation implicit in teaching modern mathematics with an abacus. This inadequation is not *modern,* but it is exposed today better than ever before." Jacques Derrida, *Of Grammatology,* trans. Gayatri Chakravorty Spivak, Corrected Edition (Baltimore, MD: Johns Hopkins University Press, 1974), 86–87.

81. Pound, *Guide to Kulchur,* 121. "The language of prose is much less highly charged, that is perhaps the only availing distinction between prose and poesy. Prose permits greater factual presentation, explicitness, but a much greater amount of language is needed." Ezra Pound, "How to Read," in *Polite Essays* (Freeport, NY: Books for Libraries Press, 1966), 171. Cf. Pound's broadcast of May 21, 1942, "E. E. Cummings Revisited": "I once had a kike friend who had a theory about poetry, namely that no one ever read ALL the words on a page. It did not, that that theory did not lead him to make his poetry quite satisfactory, but as theory it may also serve [to] enlighten you" (*Ezra Pound Speaking,* 143).

82. Derrida, *Of Grammatology,* 68.

83. Pound, *Guide to Kulchur,* 57.

84. Niklas Luhmann, *Social Systems,* trans. John Bednarz Jr. with Dirk Baecker (Stanford, CA: Stanford University Press, 1995), li. In the following sections, I appropriate his category of interpenetration to identify the process whereby revelation is enabled. For Luhmann, interpenetration occurs simultaneously when "both systems enable each other by introducing their own already-constituted complexity into each other" and "the receiving system also reacts to the *structural formation* of the penetrating system" (213; emphasis in original).

85. Pound, *Guide to Kulchur,* 121.

86. Ibid., 51.

87. Ezra Pound, *The Spirit of Romance* (New York: New Directions, 1968), 14; quoted in *The Sayings of Ezra Pound,* ed. G. Singh (London: Duckworth, 1994), 11.

88. Luhmann, *Social Systems,* 232. "If one determines the schema itself, then one can

leave to the other system the choice between two possibilities. The complexity of the other system is accepted insofar as one does not know which of the two possibilities it will choose; yet that complexity is rendered unproblematic because one has ready connective behavior for both possibilities" (233).

89. Ibid., 220.

90. In this regard, see Pound's comments on religion: "A live religion cannot be maintained by scripture. It has got to go into effect repeatedly in the persons of the participants." *Guide to Kulchur,* 191.

91. Pound glosses his search as one that does not necessarily require the category of the human: "I am, I trust patently, in this book doing something different from what I attempted in *How to Read* or in the *ABC of Reading.* There I was avowedly trying to establish a series or a set of measures, standards, voltometers, here I am dealing with a heteroclite set of impressions. I trust human, without their being too bleatingly human" (ibid., 208).

92. Ibid., 295.

93. Ibid., 286.

94. Eric A. Havelock, *Preface to Plato* (Oxford: Basil Blackwell, 1963), 100. Havelock's thesis suffers, however, from a too severe epoch making. "Mimetic and diegetic orality are equally ancient, and writing, for that matter, has been in use for millennia. Oral forms of artistic expression did not begin to wither away in the second millennium BC with the invention of alphabetic script or in the second millennium AD with the invention of movable type. We find cultural clashes associated with different modes of information storage and retrieval, but we also note remarkable evidence of ongoing synergy and assimilation." Christopher Collins, *The Poetics of the Mind's Eye* (Philadelphia: University of Pennsylvania Press, 1991), 32.

95. Pound, *Guide to Kulchur,* 51.

96. Derrida, *Of Grammotology,* 333.

97. "In general the new demotes the old. Each mediological revolution gives rise to its own producers of friction, its own 'switch points' along the tracks of its development, and the history of a cultural milieu can be read (and written) as that of the short-circuits and competitions between juxtaposed, or rather, superposed, devices of transmission." Régis Debray, *Media Manifestos: On the Technological Transmission of Cultural Forms,* trans. Eric Rauth (London: Verso, 1996), 16.

98. Pound, *Guide to Kulchur,* 291.

99. Ibid., 317.

100. "An author uses a certain number of blank words for the timing, the movement, etc., to make his work sound like natural speech. I believe one should check up all that verbiage as say 4% blanks, to be used where and when wanted in translation." Ezra Pound, Rapallo, to W. H. D., undated, *The Letters of Ezra Pound, 1907–1941,* ed. D. D. Paige (London: Faber and Faber, 1951), 357–58.

101. Doob, introduction to *Ezra Pound Speaking,* xi. "By January 1941, he was recording talks for broadcast on Rome Radio's American Hour, at 350 lire a time, with Vivaldi's recordings, by his stipulation, played before and after. He thought the Vivaldi's should be heard in America, and enjoyed them himself, on his couch in Rapallo, listening to

playbacks. He found his voice strange." Hugh Kenner, *The Pound Era* (London: Faber & Faber, 1972), 465.

102. Pound was aware of the problems presented by atmospheric interference. "In between a fine clear and strong Berlin on one edge, and B.B.C. nuisance on the other / Only resolute determination to get Rome would have led anyone to it. It comes on 29.6 on my dial, not at 31 as announced." Ezra Pound, Rapallo, to Ungaro, April 22, 1942; quoted in Tim Redman, *Ezra Pound and Italian Fascism* (Cambridge: Cambridge University Press, 1991), 221. There are an enormous number of biographies and critical articles describing Pound's broadcasts and subsequent arrest. For the crucial primary documents, see the anthology *Ezra and Dorothy Pound: Letters in Captivity, 1945–46,* ed. Robert Spoo and Omar Pound; Redman's study, which offers a discussion of how Pound came to broadcasting; and James J. Wilhelm, *Ezra Pound: The Tragic Years, 1925–1972* (University Park: Pennsylvania State University Press, 1994).

103. De Rachewiltz, "Fragments," 161.

104. Ibid., 162. The letter to Duncan reads: "Blasted friends left a goddam radio here yester. Gift. God damn destructive and dispersive devil of an invention. But got to be faced. Drammer has got to face it, not only face cinema. Anybody who can survive may strengthen inner life, but mass of apes and worms will be still further rejuiced to passivity. Hell a state of passivity? Or limbo? Anyhow what drammer and teeyater wuz, radio is." *Letters of Ezra Pound,* 441.

105. Ezra Pound, Rapallo, to Ungaro, June 27, 1941; quoted in Redman, *Ezra Pound and Italian Fascism,* 211.

106. "Sworn Statement by Ezra Pound to the Office of the Counter Intelligence Corps of the U.S. 92nd Army," in *Ezra and Dorothy Pound: Letters in Captivity,* 61.

107. Ibid., 63. This is at odds with G. Fuller Torrey's account of Pound's broadcasting career. Torrey has Pound contributing short news items and comments, "many anti-Semitic in nature," to *Jenny's Front Calling,* a program in which captured American soldiers were identified by name and messages were sent to their relations in the States. "Pound also created a character called 'Mr. Dooley' who made humourous remarks about the Allied war effort." In *The Roots of Treason: Ezra Pound and the Secret of St. Elizabeth's* (New York: McGraw-Hill, 1984), 173. Torrey's account is, by most accounts, tendentious. See instead Tim Redman's remarkable unearthing of Pound's writings for the Salò Republic. His conclusion? "Pound apparently continued some kind of collaboration with radio propaganda during this time, though details of it are very difficult to obtain." Redman, *Ezra Pound and Italian Fascism,* 262.

108. Charles Norman, *The Case of Ezra Pound* (New York: Funk and Wagnalls, 1968), 56.

109. *Ezra and Dorothy Pound: Letters in Captivity,* 61.

110. "The odd jargon and mixture of dialects that he employed had convinced the Fascist secret service, which was prone to error as its American counterpart, that he was sending messages in code to the United States armed forces!" Eustace Mullins, *This Difficult Individual Ezra Pound* (New York: Fleet Publishing, 1961), 203.

111. De Rachewiltz, "Fragments," 165.

112. Ibid., 161.

113. Friedrich Kittler, "Benn's Poetry—'A Hit in the Charts': Song under Conditions of Media Technologies," *Substance: A Review of Literary Theory and Criticism,* no. 61 (1990): 18.

114. "That is the disadvantage of the radio form, and heaven knows when I shall be able to print these texts in book or books available to the American and English public." "Continuity," July 6, 1942, *Ezra Pound Speaking,* 191.

115. De Rachewiltz, "Fragments," 164. Unfortunately, it is unclear what text Ms. de Rachewiltz is citing. For her part, Dorothy Pound says much the same in a letter to McLuhan: "E. P. is more interested in agenda, than in analysis of the past. To get a few of you scholars to combine and break the deadlock of all live scholarship; improve the curriculum by definitely insisting on a better set of 50 (or even 100) books . . . He says some, indeed, most of your questions are answered in the 80 cantos." Dorothy Pound, to Marshall McLuhan, June 21, 1948; quoted in Barton, "On the Ezra Pound Marshall McLuhan Correspondence."

116. Doob, introduction to *Ezra Pound Speaking,* xi. Pound had been assiduously collecting data for years. See his speech "Consolidate" from 1942: "A considerable force, in some cases a force of inertia, has been espoused to my views, to my perceptions, to my patient collecting of data, ever since I had any views or perceptions, or started collecting data" (394).

117. Martin Van Creveld, *Command in War* (Cambridge, MA: Harvard University Press, 1985), 7.

118. Kittler, *Gramophone, Film, Typewriter,* 193.

119. Norman, *The Case of Ezra Pound,* 160.

120. The word "typewriter" "originally referred both to the machine and its operator, usually a young woman." Mark Seltzer, "The Graphic Unconscious: A Response," *New Literary History* 26 (1995): 24. "'Typewriter,' after all signifies both: machine and woman. Two years after the purchase of the machine, Freud wrote to Abraham from Hofmannsthal's Vienna: 'A quarter of an hour ago I concluded the work on melancholy. I will have it typewritten so that I can send you a copy.'" Kittler, *Gramophone, Film, Typewriter,* 216.

121. De Rachewiltz, *Discretions,* 150.

122. On poetry and learning by heart, see Jacques Derrida, "Che cos'è la poesia," in *The Derrida Reader,* trans. Peggy Kamuf (New York: Columbia University Press, 1991), 289–99.

123. De Rachewiltz, *Discretions,* 128.

124. Consider her fascination with oracles: "Long before I had been fascinated by oracles, doom, hubris; before I had been made aware of and warned against Freudian trends and the dangers of oversimplification." De Rachewiltz, *Discretions,* 189.

125. Ibid., 3.

126. Ibid., 155.

127. Ibid., 189.

128. Ibid., 190.

129. Ezra Pound, "George Antheil *(Retrospect),*" in *Ezra Pound and Music: The Complete Criticism* (London: Faber & Faber, 1977), 257. In addition, see Pound's postscript:

"It may be necessary to fill in the gaps between the wires (the sacrosanct 1sts, 3ds, and 5th) even if it cause pain to some ears avid of succulence and insensitive to the major form" (265).

130. De Rachewiltz, *Discretions*, 118.

131. Ibid., 139.

132. Ibid., 156. Pound's cuts via the secretarial circuit bring to mind the logic of anorexia, the subject of Leslie Heywood's reading of Pound and Eliot. See in particular the relation she draws between female padding, secretaries, and the phallic stylus: "This is the position of the male anorexic, who eliminates female padding on the way to phallic creativity. For Pound and Eliot, the feminine needs to be cut, cut up, burned away to arrive at art. They affirm a return to an originary purity, some time or place that was 'better' because more masculine and clear, less chaotic, where the poet had more control in a vitally religious world." Leslie Heywood, *Dedication to Hunger: The Anorexic Aesthetic in Modern Culture* (Berkeley: University of California Press, 1996), 219. On this note, see de Rachewiltz's losing battle with peanut butter: "I stopped eating peanut butter and writing poems" (*Discretions*, 262).

133. Ibid., 159.

134. Recalling Pound's supposed attempts to leave Italy upon its declaration of war with the United States, de Rachewiltz writes: "I do not remember details. The words that stick in my mind are: clipper, the last clipper, frozen assets, frozen bank account. Grandfather's U.S. government pension withheld, the old man in the hospital with a broken hip" (*Discretions*, 152).

135. Ezra Pound, *If This Be Treason* (Siena: Tip Nuova, 1948), 4.

136. Telephone operator or secretary, women are essential for acts of dictation. "Women alone could commit the act of speech dictated since 1881, which combines a third person with a deixis and is, according to Benveniste, completely impossible: 'Switchboard here, what is required?'" Bernhard Siegert, "Switchboards and Sex: The Nut(t) Case," in *Inscribing Science: Scientific Texts and the Materiality of Communication*, ed. Timothy Lenoir (Stanford, CA: Stanford University Press, 1998), 85.

137. Norman, *The Case of Ezra Pound*, 5.

138. Noel Stock, *The Life of Ezra Pound* (New York: Avon Books, 1970), 508.

139. Kittler, *Discourse Networks*, 330. See Doob's description of Pound's manuscripts: "Elementary misspelling has been corrected. Punctuation and paragraphing have been altered in the interest of intelligibility" (*Ezra Pound Speaking*, xiii).

140. "Zion," in *Ezra Pound Speaking*, 242. For the relevant discussions on poetics and anti-Semitism, see Anthony Julius, *Anti-Semitism and Literary Form* (New York: Cambridge University Press, 1995); and Robert Casillo, *The Genealogy of Demons: Anti-Semitism, Fascism, and the Myths of Ezra Pound* (Evanston, IL: Northwestern University Press, 1988).

141. Pound, *Ezra Pound Speaking*, 190.

142. Ibid.

143. Ibid., 192.

144. Ibid., 194.

145. See the speech of February 12, 1942, in which Pound reads Canto 46, or his

reflections on broadcasting from London and Berlin in his talk from March 7, 1943. Addressing himself to the United Kingdom, he writes: "The monotony of your evasions breeds infinite boredom. Berlin by contrast is placid, as against your gallic hysteria. Patient but firm German voices go on explaining *urbe et orbe* just what the war is about" (*Ezra Pound Speaking,* 236).

146. Ibid.

147. Ezra Pound, London, to Homer Pound, August 23, 1917; quoted in "Fragments," 158.

148. Or had an air-driven, wireless typewriter. "Normally, the air in the rotating vane escapes through the keys by way of these openings; but when a key is covered by the finger and the vane in its circuit passes the closed tube, the flow of air through the vane is momentarily checked, and instantly the type . . . is thrown by the diaphragm against the plates. The character is printed in less than 1/120 of a second. . . . The type wheel thus practically moves constantly. It will be readily understood that in certain cases a number of keys may be touched at the same time, and the various characters will be printed in rapid succession. By a proper arrangement of the keyboard one of the shorter words and common syllables may thus be printed at a single operation." "An Air-Driven Typewriter," *Scientific American,* no. 1707 (September 19, 1908): 181.

149. Pound, *Ezra Pound Speaking,* 192. "The man who had such trouble, and at his leisure, in focusing the Monte dei Paschi Canto clearly, was still less successful, under these conditions, in manifesting a semblance of calm. Exacerbations needled him." Hugh Kenner, *The Pound Era* (Berkeley: University of California Press, 1971), 466.

150. Pound, *Guide to Kulchur,* 53.

151. Marshall McLuhan, "Effects of the Improvements of Communication Media," *Journal of Economic History* 20 (December 1960): 575. It is difficult to imagine a more Poundian moment on the part of McLuhan than this. For more of McLuhan on Pound, see Marshall McLuhan, "Pound's Critical Prose," *The Interior Landscape: The Literary Criticism of Marshall McLuhan* (New York: McGraw-Hill, 1969), 75–81.

152. Pound, *Ezra Pound Speaking,* 191.

153. C. A. Siepmann, "Can Radio Educate?" *Journal of Educational Sociology* 14 (February 1941): 346.

154. Alan Williams, "Is Sound Recording Like a Language?" *Yale French Studies,* no. 60 (1980): 57.

155. Lawrence Rainey, "Taking Dictation: Collage Poetics, Pathology, and Politics," *Modernism/Modernity* 5 (1998): 148.

156. Ibid.

157. *Mussolini* (Zeus Editoriale, 1996), 125.

158. Arnheim, *Radio,* 14.

INDEX

Timothy C. Campbell is assistant professor of Italian at Cornell University. He has written on modernism and radio, Italian Futurism, and the role of technology in fascist culture.